Applied Cartesian Tensors for Aerospace Simulations

Applied Cartesian Tensors for Aerospace Simulations

David M. Henderson
Lago Vista, Texas

EDUCATION SERIES
Joseph A. Schetz
Series Editor-in-Chief
Virginia Polytechnic Institute and State University
Blacksburg, Virginia

Published by
American Institute of Aeronautics and Astronautics, Inc.
1801 Alexander Bell Drive, Reston, VA 20191-4344

American Institute of Aeronautics and Astronautics, Inc., Reston, Virginia

1 2 3 4 5

Library of Congress Cataloging-in-Publication Data

Henderson, David M. (David Melvin), 1935–
Applied cartesian tensors for aerospace simulations/David M. Henderson.
p. cm. — (AIAA education series)
Includes bibliographical references and index.
ISBN 1-56347-793-9

1. Flight simulators—Mathematics. 2. Calculus of tensors. I. Title. II. Series.

TL712.5.H46 2006
629.132'52078—dc22 2006000085

Foreword

We are pleased to present *Applied Cartesian Tensors for Aerospace Simulations* by David M. Henderson. This compact volume covers the use of tensors in the analysis of navigation, guidance, and control of aerospace vehicles. There are four chapters and four appendices, and the book is more than 200 pages long. David Henderson is very well qualified to write on the subject, because of his long and broad experience in the field.

The AIAA Education Book Series aims to cover a very broad range of topics in the general aerospace field, including basic theory, applications and design. Information about the complete list of titles can be found on the last pages of this volume. The philosophy of the series is to develop textbooks that can be used in a university setting, as instructional materials for continuing education and professional development courses, and also as books that can serve as the basis for independent study. Suggestions for new topics or authors are always welcome.

Joseph A. Schetz
Editor-in-Chief
AIAA Education Book Series

Table of Contents

Preface

This book was written for aerospace students who have special interests in the mathematical description of the motion of flight vehicles. The book has been written from collections of mathematical derivations, notes, and technical reports. The material covers a wide range of aerospace disciplines and has been taken from the work of my colleagues as well as my work, which has been directly related to the solution of problems in navigation, guidance, and control of aerospace vehicles. The book can serve as both a supplemental text or an instructive book covering the fundamental concepts of the geometry of space, applied mechanics, and aerospace engineering analysis. I feel that understanding these concepts is essential in flight mechanics to describe the motion of aircraft and space vehicles. For this reason, I have included selected aerospace application topics at the end of each section. These subsections use some parts of the analysis that was presented and demonstrate its application as used today in actual navigation, guidance, and control software functions.

The reader should have an understanding of differential and integral calculus, differential equations, and the fundamentals of vector analysis. The basic concepts of Cartesian analysis are developed along with the application of tensor notation to engineering analysis. Tensor notation (the Einstein summation convention) is introduced to give the reader exact component equations and to demonstrate its value in multivariable analysis. Mathematical terms that are represented using the summation notation may not display tensor character, but the notation is used for its power and clarity to present exact component definitions and for brevity of expression.

Derivations of fundamental equations are presented using the summation convention, which suggests a sightly different mathematical logic in comparison with vector-matrix notation. The use of this notation, in some cases, provides for more general solutions and hopefully a foundation leading to better understanding of aerospace problems of the future. The advantage of this method of analysis and this notation is that each equation represents specific physical components in the desired coordinate system. Also the analysis results using this notation lend themselves to direct computer applications.

Concise definitions of fundamental measurements are presented, and care has been taken to use only variables in the right-hand side of equations that have been previously defined or have appeared on the left-hand side in a defining relationship. Topics are arranged in logical progression beginning with the fundamental transformation equations in the absence of mass and energy. The basic concepts of rotating coordinate systems lead to the equations of motion for aircraft and space vehicles in gravitational fields including, for instance, the effects of gravitational torques. Operational use of quaternions for vehicle attitude determination as applied to navigation, guidance, and control analysis is presented. Chapter 4 is devoted to the analysis required to simulate the motion of actual aerospace vehicles. Specific solutions to the equations of motion for flight vehicles are not discussed, however, detailed numerical methods are described showing how these differential equations are solved.

The text is written purposely without direct references to specific vehicles or aerospace projects. The intent of the author is to describe the analysis in a general manner so that the work can be applied to motion of any aerospace vehicle. Some of the topics, especially those related to finite element aerodynamic analysis and the special solutions of the two-body problem, are discussed with only a brief outline of the pertinent details of the mathematical analysis. I feel that the abbreviated inclusion of these analyses are important for educational purposes and for the overall continuity of the text, but the detailed mathematical analysis was beyond the scope of this text. However, I do provide the reader with generous references to the many fine texts covering those topics in much greater detail.

I feel that, by applying the summation notation in the analysis, a more complete description of the dynamic problems of aerospace vehicle motion can be offered. These concepts and this type of analysis, and some results from general relativity, are now finding applications in aerospace engineering technologies. Hopefully, the text gives some new insights to the solutions of old problems in vehicle dynamics and provides the reader with a clearer foundation and a heightened curiosity for the solutions of new problems in the future.

David M. Henderson
December 2005

Acknowledgments

During my aerospace career, I have worked with many engineers and scientists and have learned greatly from them. I hope that the explanatory style of the text presents the analyses and information concisely for the young engineers and scientists coming into this field of knowledge.

I would like to acknowledge many of those who worked with me over the years, but they are too many for me to recognize here. They shared much of their analysis experience with me, which I have used in the text. I have tried to give them due reference when their work was used in the text. Again, I have gained diverse experiences from the aerospace communities that have given me the material for this collection of concepts fundamental for the analysis of the motion of aircraft and space vehicles.

I would like to express my special thanks in memory of Marie Holmes, who typed the original draft of Chapter 1. Also, special thanks go to Mike Fraietta, who provided me with expert technical assistance in solving computer system problems and just plain "making things work". Mike's wealth of aerospace engineering and analytic abilities has also provided me with many solution methods and has been a source of reference materials for the text. I am grateful to Mike for proofreading the text and offering welcome suggestions for changes and corrections to the manuscript. I am also grateful to Larry Frieson and Al Jackson, both of whom over the years have taught me how to be a scientist. They have also worked to proofread and critique the text. Mark Allman was the guiding inspiration for the LaTex program and generously provided his expertise in that system. He also helped in the preparation and proofreading of the text for publication.

I am also very grateful to my wife Elaine, who has been helpful in the preparation of the manuscripts and for understanding the reasons for my work hours over these last years.

1
Geometric Concepts in the Absence of Mass and Gravitation

Introduction

In the first parts of this chapter I have tried to bridge a gap between physics and engineering for the aerospace student by showing the significance of the tensor to engineering analysis. The Cartesian position transformation matrix equation is derived from the geometric equivalence of a radius vector as viewed from two different coordinate frames. The transformation equation is then written using the Einstein summation convention and is shown to be a tensor equation displaying the same form as the general tensor equation from mathematical physics. The tensor equation is the law of transformation for the tensor components themselves.

Using the properties of the Cartesian transformation matrix, for example, transformation matrices are orthogonal, many special relationships involving the matrix elements are developed. The algebraic identities from these properitie are used in analysis for the reduction of many aerospace defining equations.

Representing the transformation matrix as a function of three Euler angles is the most used method and certainly the most understandable by the human mind. There are twelve different sets of Euler angle sequences using three Euler angles which can be used to represent a given transformation matrix, all of which are equally as accurate in defining the matrix. However, it is shown that a specific Euler sequence set may be better applied to the analysis of a particular problem.

Representing the transformation matrix as a function of the four parameter methods, such as the Hamilton quaternion, is mathematically better, especially when simulating motion dynamics. However, these methods are harder to visualize in the mind, but the computer software systems do not have this problem. The necessity to use the four parameter methods is driven by the inherent singularity condition, which occurs as one of the Euler angles passes through 90 degrees. This singularity can occur in each of the Euler sequences used to define the three Euler angles. In many aerospace problems and in this text, the use of the Hamilton quaternion has become the recommended standard method for defining the transformation matrix.

Defining and understanding how the transformation matrix varies with time is an important relationship. The time derivative of the transformation matrix can be expressed as functions of the Euler angle rates, of change, the coordinate axis rotation rates, or the quaternion rates of change. Selecting the best expression, with its parameter rates of change, is determined by the analysis required in the particular problem.

The general forms for the transformation equations for the velocity and acceleration vectors are derived by differentiation of the position transformation equation. These reference transformation equations are used in the development of many aerospace dynamic equations of motion.

A subsection of aerospace applications is presented at the end of each section. These subsections contain examples of simulation methods which are in use today describing real guidance navigational control (GNC) software functions. For a simple example, again using the fundamental properties of the transformation matrix, it is shown how the onboard computer software computes the Local-Vertical-Local-Horizontal to inertial transformation matrix using the best estimates for the navigated inertial position and velocity vector states. This matrix may then be used as required input to other onboard software functions.

1.1 Position Transformation

1.1.1 Direction Cosine Matrix

We can construct a coordinate system at some point in space and for some region about the point, by selecting three coordinate axes such that they are perpendicular to one another. The three coordinate axes are defined in a special manner using the *right-hand rule*: with the extended thumb of the right hand, the curled fingers show the rotation direction to the next axis in the plane perpendicular to the pointing thumb. Each axis is defined this way, thus the coordinate system is called a right-handed system. A point P within this region may be located by three physical measurements, one along each axis, x_1, x_2, x_3 at some time x_4. This notation is precisely the same as using x, y, z at some time t. These are the fundamental assumptions for a three-dimensional Cartesian coordinate system in Euclidean space. The distance r to the point P can be computed by

$$r^2 = x_1^2 + x_2^2 + x_3^2 \text{ at some time } x_4 \tag{1.1}$$

The variable x_4 represents the time when the physical measurements are taken. Time is the universal independent parameter and is an integral part of the definition of space itself. However, in Euclidean space, the geometry of position location is not a function of time, except of course when motion occurs over some interval of time. In vector notation the radius vector to a point in the coordinate system may be represented as follows:

$$\boldsymbol{r} = x_1\boldsymbol{i} + x_2\boldsymbol{j} + x_3\boldsymbol{k} \tag{1.2}$$

where $\boldsymbol{i}, \boldsymbol{j}$, and $\boldsymbol{k}$ are mutually perpendicular unit vectors and x_1, x_2, and x_3 are the magnitudes of the components of the vector $\boldsymbol{r}$ along each coordinate axis. We further assume a second Cartesian coordinate system that has been translated and rotated from the orginal reference frame. The original reference coordinate system will be referred to as the inertial or stationary frame. The second frame will be

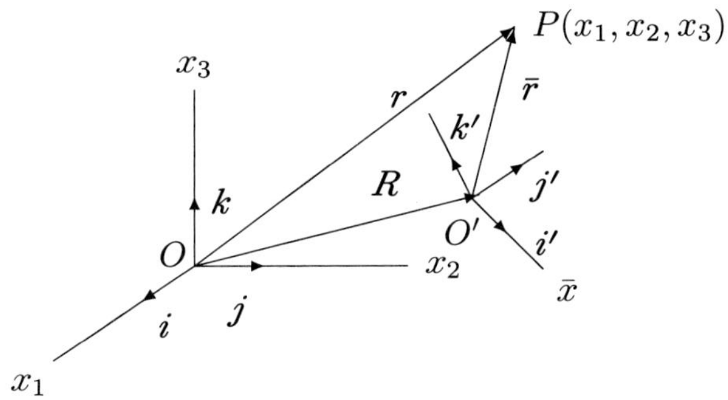

Fig. 1.1 Cartesian coordinate frames.

referred to as the moving frame. Then at some instant in time we can write

$$\boldsymbol{r} = \boldsymbol{R} + \bar{\boldsymbol{r}} \tag{1.3}$$

This is illustrated in Fig. 1.1.

Also shown in Fig. 1.1 are $\boldsymbol{i}'$, $\boldsymbol{j}'$, and $\boldsymbol{k}'$ which are unit vectors mutually perpendicular at orgin O'. The magnitudes of the components of the vector $\bar{\boldsymbol{r}}$ are $\bar{x}_1$, $\bar{x}_2$, and $\bar{x}_3$ as measured along the axes constructed at O'. Rewriting Eq. (1.3) for clarity, we have

$$\bar{\boldsymbol{r}} = \boldsymbol{r} - \boldsymbol{R} \tag{1.4}$$

If O' is located at X_1, X_2, and X_3 in the reference coordinate system at orgin O, we can write

$$\boldsymbol{R} = X_1\boldsymbol{i} + X_2\boldsymbol{j} + X_3\boldsymbol{k} \tag{1.5}$$

Writing Eq. (1.4) in component form,

$$\bar{\boldsymbol{r}} = (x_1 - X_1)\boldsymbol{i} + (x_2 - X_2)\boldsymbol{j} + (x_3 - X_3)\boldsymbol{k} \tag{1.6}$$

where $\boldsymbol{r}$ is seen from origin O. However, $\bar{\boldsymbol{r}}$ as seen from O' is simply

$$\bar{\boldsymbol{r}} = \bar{x}_1\boldsymbol{i}' + \bar{x}_2\boldsymbol{j}' + \bar{x}_3\boldsymbol{k}' \tag{1.7}$$

Since we are in Euclidean space, $\bar{\boldsymbol{r}}$ is an invariant as seen from either coordinate system O or O', therefore the vector $\bar{\boldsymbol{r}}$ from Eq. (1.6) must be identical to $\bar{\boldsymbol{r}}$ of Eq. (1.7), and we can write

$$(x_1 - X_1)\boldsymbol{i} + (x_2 - X_2)\boldsymbol{j} + (x_3 - X_3)\boldsymbol{k} = \bar{x}_1\boldsymbol{i}' + \bar{x}_2\boldsymbol{j}' + \bar{x}_3\boldsymbol{k}' \tag{1.8}$$

If we form the vector dot product of Eq. (1.8) with $\boldsymbol{i}, \boldsymbol{j}$, and $\boldsymbol{k}$, respectively, one at a time, we have

$$\begin{aligned}
x_1 - X_1 &= (\boldsymbol{i}\cdot\boldsymbol{i}')\bar{x}_1 + (\boldsymbol{i}\cdot\boldsymbol{j}')\bar{x}_2 + (\boldsymbol{i}\cdot\boldsymbol{k}')\bar{x}_3 \\
x_2 - X_2 &= (\boldsymbol{j}\cdot\boldsymbol{i}')\bar{x}_1 + (\boldsymbol{j}\cdot\boldsymbol{j}')\bar{x}_2 + (\boldsymbol{j}\cdot\boldsymbol{k}')\bar{x}_3 \\
x_3 - X_3 &= (\boldsymbol{k}\cdot\boldsymbol{i}')\bar{x}_1 + (\boldsymbol{k}\cdot\boldsymbol{j}')\bar{x}_2 + (\boldsymbol{k}\cdot\boldsymbol{k}')\bar{x}_3
\end{aligned} \tag{1.9}$$

Likewise, forming the dot products of Eq. (1.8) with $\boldsymbol{i}', \boldsymbol{j}'$, and $\boldsymbol{k}'$, respectively,

$$\bar{x}_1 = (x_1 - X_1)(\boldsymbol{i}\cdot\boldsymbol{i}') + (x_2 - X_2)(\boldsymbol{j}\cdot\boldsymbol{i}') + (x_3 - X_3)(\boldsymbol{k}\cdot\boldsymbol{i}')$$
$$\bar{x}_2 = (x_1 - X_1)(\boldsymbol{i}\cdot\boldsymbol{j}') + (x_2 - X_2)(\boldsymbol{j}\cdot\boldsymbol{j}') + (x_3 + X_3)(\boldsymbol{k}\cdot\boldsymbol{j}')$$
$$\bar{x}_3 = (x_1 - X_1)(\boldsymbol{i}\cdot\boldsymbol{k}') + (x_2 - X_2)(\boldsymbol{j}\cdot\boldsymbol{k}') + (x_3 - X_3)(\boldsymbol{k}\cdot\boldsymbol{k}') \tag{1.10}$$

Eqs. (1.9) and (1.10) transform the position vector of a point in one system to the position vector of that same point as viewed from the other coordinate system. Remembering that $(\boldsymbol{i}\cdot\boldsymbol{i}')$ = cos(angle between $\boldsymbol{i}$ and $\boldsymbol{i}'$), we can let each cosine angle be

$$a_1^1 = (\boldsymbol{i}\cdot\boldsymbol{i}'),\quad a_1^2 = (\boldsymbol{i}\cdot\boldsymbol{j}'),\quad a_1^3 = (\boldsymbol{i}\cdot\boldsymbol{k}')$$
$$a_2^1 = (\boldsymbol{j}\cdot\boldsymbol{i}'),\quad a_2^2 = (\boldsymbol{j}\cdot\boldsymbol{j}'),\quad a_2^3 = (\boldsymbol{j}\cdot\boldsymbol{k}')$$
$$a_3^1 = (\boldsymbol{k}\cdot\boldsymbol{i}'),\quad a_3^2 = (\boldsymbol{k}\cdot\boldsymbol{j}'),\quad a_3^3 = (\boldsymbol{k}\cdot\boldsymbol{k}') \tag{1.11}$$

This will simplify our notation, and Eqs. (1.9) can be written in component form as

$$(x_i - X_i) = \sum_{\alpha=1}^{3} a_i^\alpha \bar{x}_\alpha \tag{1.12}$$

and Eqs. (1.10) will take the form

$$\bar{x}^i = \sum_{\alpha=1}^{3} a_\alpha^i (x^\alpha - X^\alpha) \tag{1.13}$$

where the superscripts are not to be confused with exponentiation, and in Cartesian analysis, as it will be discussed, we have $x^\alpha = x_\alpha$, $\bar{x}^\alpha = \bar{x}_\alpha$, and $X^\alpha = X_\alpha$. Note that the small letter i is used to represent the ith measurement along the ith axis. Also notice that each of the a_i^α of Eq. (1.11) is the cosine of the angle between the indicated axes.

1.1.2 Einstein Summation Convention and Introduction to Cartesian Tensors

We can go one step further to simplify our notation by using the summation convention that Einstein introduced to tensor calculus in 1916 (see Ref. 1, p. 122, and Ref. 2, Sec. 1.2). Equations (1.12) and (1.13) can be written as follows:

$$x_i - X_i = a_i^\alpha \bar{x}_\alpha \tag{1.14}$$

and

$$\bar{x}^i = a_\alpha^i (x^\alpha - X^\alpha) \tag{1.15}$$

in standard summation notation. Einstein once wrote (Ref. 2, p. 21), "*I made a great discovery in mathematics; I suppressed the summation sign every time that the summation has to be done on an index which appears twice in the general term.*" Here, since we are in a three-dimensional space, all summations are automatically assumed to go from one to three whenever repeated indices occur in the same equation. In this text the Greek letters will be used to indicate summations, and the small letters such as i, j, and k will be used to identify the component axes. The summation notation will be used throughout the text to represent the specific components defining each of the terms in the equation.

Equations such as (1.14) and (1.15) are called *Cartesian tensor equations* (see Ref. 3, pp. 33, 34), since they are the results of orthogonal Cartesian transformations. Tensor equations have many interesting qualities and define the specific mathematical laws for coordinate transformations of the variables involved. The $(x_i - X_i)$ and $\bar{x}_i$ are called first-order tensors forming the components of the position vectors.

If we take the partial derivatives of Eqs. (1.14) and (1.15), we have

$$\left(\frac{\partial x_i}{\partial \bar{x}_\alpha}\right) = a_i^\alpha \tag{1.16}$$

and

$$\left(\frac{\partial \bar{x}_i}{\partial x_\alpha}\right) = a_\alpha^i \tag{1.17}$$

respectively. Then Eqs. (1.14) and (1.15) could also be written

$$x_i - X_i = \left(\frac{\partial x_i}{\partial \bar{x}_\alpha}\right) \bar{x}^\alpha \tag{1.18}$$

and

$$\bar{x}_i = \left(\frac{\partial \bar{x}_i}{\partial x_\alpha}\right) (x^\alpha - X^\alpha) \tag{1.19}$$

The vectors $(\boldsymbol{x} - \boldsymbol{X})$ and $\bar{\boldsymbol{x}}$ whose coordinate transformation forms are defined like Eqs. (1.18) and (1.19) are called *first-order contravariant tensors*, which are the components of *contravariant vectors* (see Ref. 4, p. 271). The analysis used in Eq. (1.8) shows the invariant quality of the length of the contravariant vector.

Cartesian tensors are defined based upon $a_i^\alpha = a_\alpha^i$, i.e., the inverse of the transformation matrix is equal to its transpose, and hence from Eqs. (1.16) and (1.17), we have the following series of equalities:

$$\frac{\partial x_i}{\partial \bar{x}_\alpha} = \frac{\partial x_\alpha}{\partial \bar{x}_i} = \frac{\partial \bar{x}_i}{\partial x_\alpha} = \frac{\partial \bar{x}_\alpha}{\partial x_i} \tag{1.20}$$

The special properties of the transformation matrix are discussed in detail in Sec. 1.2 of the text. Based on Eq. (1.20), Eqs. (1.18) and (1.19) could also be written

$$x_i - X_i = \left(\frac{\partial \bar{x}_\alpha}{\partial x_i}\right) \bar{x}^\alpha \tag{1.21}$$

and

$$\bar{x}_i = \left(\frac{\partial x^\alpha}{\partial \bar{x}_i}\right)(x_\alpha - X_\alpha) \tag{1.22}$$

The tensors that display the general form of Eqs. (1.21) and (1.22) are called first-order covariant tensors and form the components of *covariant vectors*, which is also invariant by coordinate transformations. From this discussion it can be seen that in Cartesian analysis *transformations affect contravariant vectors in exactly the same way as they do covariant vectors*. Therefore, there is no distinction made between covariant and contravariant vectors in Cartesian analysis.

The preceding mathematical forms belong to the more general differential invariants called tensors whose components transform according to

$$\bar{T}^{j_1,j_2,\ldots,j_m}_{i_1,i_2,\ldots,i_n} = \left|\frac{\partial x}{\partial \bar{x}}\right|^N T^{\alpha_1,\alpha_2,\ldots,\alpha_m}_{\beta_1,\beta_2,\ldots,\beta_n} \frac{\partial \bar{x}^{j_1}}{\partial x^{\alpha_1}} \cdots \frac{\partial \bar{x}^{j_n}}{\partial x^{\alpha_m}} \frac{\partial x^{\beta_1}}{\partial \bar{x}^{i_1}} \cdots \frac{\partial x^{\beta_n}}{\partial \bar{x}^{i_n}} \tag{1.23}$$

where the determinant $\left|\frac{\partial x}{\partial \bar{x}}\right|$ is called the *Jacobian of the transformation*. In the general form of Eq. (1.23), the tensor is said to be covariant of order n and contravariant of order m. Note here that in Cartesian analysis the general tensor form of Eq. (1.23) is made simpler in that $\left|\frac{\partial x}{\partial \bar{x}}\right| = 1$, as will be shown in Sec. 1.2.2 of the text, and that we can select $N = 1$.

The Einstein summation convention is used throughout the text to reduce the size and complexity of the expressions. It must be pointed out here that all Cartesian tensors may not satisfy the mathematical definition for a tensor. In this text, tensor notation (summation notation) is used in equations to represent terms that may not be tensors, but the author feels that the extended use of this notation in applied analysis is important for the power and clarity that it brings to the subject.

1.1.3 Matrix Notation and Elements of the Transformation Matrix

Equation (1.14) can be written in matrix form simply as

$$\boldsymbol{x} - \boldsymbol{X} = (a)\bar{\boldsymbol{x}} \tag{1.24}$$

or

$$\boldsymbol{x} = (a)\bar{\boldsymbol{x}} + \boldsymbol{X} \tag{1.25}$$

where $\boldsymbol{x}$, $\bar{\boldsymbol{x}}$, and $\boldsymbol{X}$ must be column vectors and a is a 3×3 matrix. The component operations indicated by

$$a_i^\alpha \bar{x}_\alpha \tag{1.26}$$

formally defines the matrix multiply operation as shown in Eq. (1.24) and expanded becomes

$$\begin{aligned}
(x_1 - X_1) &= a_1^1\bar{x}_1 + a_1^2\bar{x}_2 + a_1^3\bar{x}_3 \\
(x_2 - X_2) &= a_2^1\bar{x}_1 + a_2^2\bar{x}_2 + a_2^3\bar{x}_3 \\
(x_3 - X_3) &= a_3^1\bar{x}_1 + a_3^2\bar{x}_2 + a_3^3\bar{x}_3
\end{aligned} \tag{1.27}$$

or before the multiply operation, Eq. (1.24) is

$$\begin{pmatrix} x_1 - X_1 \\ x_2 - X_2 \\ x_3 - X_3 \end{pmatrix} = \begin{pmatrix} a_1^1 & a_1^2 & a_1^3 \\ a_2^1 & a_2^2 & a_2^3 \\ a_3^1 & a_3^2 & a_3^3 \end{pmatrix} \begin{pmatrix} \bar{x}_1 \\ \bar{x}_2 \\ \bar{x}_3 \end{pmatrix} \tag{1.28}$$

To further standardize the Cartesian notation used in the text, the author defines the correspondence with the ith rows and jth columns used in matrix notation as follows:

$$a_i^j = a_{ij} = a^{ij}$$

or

$$a_{\text{(rows)}}^{\text{(columns)}} = a_{\text{(rows)(columns)}} = a^{\text{(rows)(columns)}} \tag{1.29}$$

Hence, Eq. (1.14) could also be written

$$x_i - X_i = a_{i\alpha}\bar{x}^{\alpha}$$

or

$$x^i - X^i = a^{i\alpha}\bar{x}_{\alpha} \tag{1.30}$$

and Eq. (1.15) could be written

$$\bar{x}^i = a^{\alpha i}(x_{\alpha} - X_{\alpha})$$

or

$$\bar{x}_i = a_{\alpha i}(x^{\alpha} - X^{\alpha}) \tag{1.31}$$

It is interesting to note the meaning of each column of the transformation matrix a with the following analysis: If we allow the coordinate centers to coincide, i.e., let all $X_i = 0$ and let the vector $\bar{\boldsymbol{x}}$ be unit vectors $(1, 0, 0)$, $(0, 1, 0)$, and $(0, 0, 1)$, one at a time, from Eq. (1.28) we have

$$\bar{\boldsymbol{x}}_1(\text{axis}) = \begin{pmatrix} a_1^1 \\ a_2^1 \\ a_3^1 \end{pmatrix}$$

$$\bar{\boldsymbol{x}}_2(\text{axis}) = \begin{pmatrix} a_1^2 \\ a_2^2 \\ a_3^2 \end{pmatrix}$$

$$\bar{\boldsymbol{x}}_3(\text{axis}) = \begin{pmatrix} a_1^3 \\ a_2^3 \\ a_3^3 \end{pmatrix} \tag{1.32}$$

Hence, each column of the transformation matrix a is a unit vector of each coordinate axis of the coordinate frame at O'. Notice that the components of the unit vector axes in Eq. (1.32) are in the coordinate frame at O. The coordinate frame at O is referred to as the *inertial or stationary reference frame.* The coordinate frame at O' is referred to as the *moving frame.*

1.1.4 Astrodynamics Application: Local-Vertical, Local-Horizontal Coordinate Reference Frame

The objective of this application is to provide the space traveler or the aerospace engineer with an understandable reference coordinate frame. The coordinate axes will be made to coincide with directions that can be readily visualized by out-the-window observations, namely the vertically downward direction and the velocity direction of the space vehicle.

We start by defining coordinate reference frames such as those in Fig. 1.1, where the center of the coordinates at O becomes the center of the Earth. This coordinate system moves with the Earth in its path through space. Further, we define the x_1 axis to point to a point fixed in space relative to the distant stars. This point that we have selected is called the vernal equinox (Ref. 5, p. M12), depicted here by the symbol Ω, which will be discussed in detail later in the text. The x_3 axis coincides with the Earth's axis of rotation. We call this frame an *inertial reference frame*, which has very important implications in physics. The inertial frame in this case defines the Earth-centered, inertial coordinate frame (the ECI coordinate frame) as shown in Fig. 1.2.

Imagine a space vehicle located at origin O' and that our navigation system can measure its position vector $\boldsymbol{R}$ and velocity vector $\boldsymbol{V}$ in the inertial frame at some point in time. Further imagine the space vehicle oriented generally in the direction of $\boldsymbol{V}$ with a space traveler seated in the space vehicle. Based on the discussion in Sec. 1.1.3, we can define the $\boldsymbol{k}'$ axis as a unit vector vertically downward toward the center of the Earth or in the negative $\boldsymbol{R}$ direction as follows:

$$\boldsymbol{k}' = -\frac{\boldsymbol{R}}{|\boldsymbol{R}|} \tag{1.33}$$

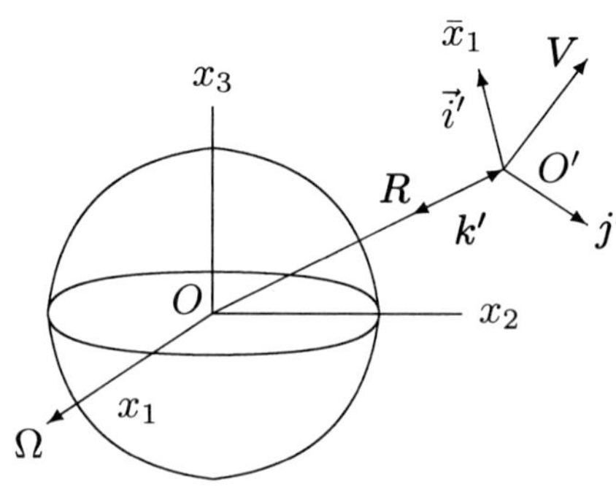

Fig. 1.2 Local-vertical, local-horizontal coordinates.

The $\boldsymbol{j}'$ axis can be formed as follows:

$$\boldsymbol{j}' = -\frac{\boldsymbol{R} \times \boldsymbol{V}}{|\boldsymbol{R} \times \boldsymbol{V}|} \tag{1.34}$$

We have attached a negative sign to the right side of Eq. (1.34) so that when the space traveler is seated, the $\boldsymbol{j}'$ axis will be in the direction of his extended right arm. The $\boldsymbol{i}'$-axis direction, which is generally in the direction of the space vehicle's motion about the Earth, is simply

$$\boldsymbol{i}' = \frac{\boldsymbol{j}' \times \boldsymbol{k}'}{|\boldsymbol{j}' \times \boldsymbol{k}'|} \tag{1.35}$$

We can now form the a matrix of Eq. (1.25), which transforms any vector in the local-vertical, local-horizontal coordinates (called the LVLH reference frame) into the ECI coordinates, by placing each of the 3×1 unit column vectors for $\boldsymbol{i}', \boldsymbol{j}'$, and $\boldsymbol{k}'$, respectively, into the 3×3 matrix,

$$(a) = (\boldsymbol{i}', \boldsymbol{j}', \boldsymbol{k}') = \left(a_\alpha^1, a_\alpha^2, a_\alpha^3\right) \tag{1.36}$$

Relative distances from the space vehicle in LVLH coordinates are measurable quantities and are used in many astrodynamic applications.

1.2 Properties of the Transformation Matrix

1.2.1 Orthogonal

The many unique properties of the transformation matrix form the mathematical foundation that is the basis for all Cartesian analysis. Understanding the coordinate transformation matrix, with the many relationships that its matrix elements have with one another and its remarkable characteristics, is important in the deviation of many of the fundamental equations in dynamics.

By examining the matrix multiply definition in Eqs. (1.25), we notice that the indicated summations on the right-hand side of Eq. (1.15) are performed on the rows of the matrix and are equivalent to $(a)^T(x - X)$ in matrix operations. Further, Eq. (1.15) can also be formed by multiplying Eq. (1.14) by $(a)^T$, which forces $(a)^T(a) = (I)$, where (I) is the unit matrix, i.e., a matrix of all zeros except for ones down the diagonal. This is the same as the definition for the matrix inverse operation $(a)^{-1}(a) = (I)$, hence the inverse of the a matrix is the same as its transpose. In matrix algebra this property of a matrix is called orthogonality (Ref. 6, p. 179). This single property of the Cartesian transformation matrix is used in the derivation of many of the basic dynamic and energy relationships of engineering analysis. Useful relationships between the elements of the a matrix, a_i^α, may be derived from the definition of the matrix inverse and the fact that a matrix and its inverse are commutative:

$$(a)^{-1}(a) = (a)(a)^{-1} = (I) \tag{1.37}$$

Conversely, if the product of two matrices form the unit matrix, then one matrix is the inverse of the other. Recall the form of the elements of the a matrix from Eq. (1.25), and since the transformation matrix is orthogonal, $(a)^{-1} = (a)^T$, we can write two matrix relations based on Eq. (1.37) as follows:

$$\begin{pmatrix} a_1^1 & a_2^1 & a_3^1 \\ a_1^2 & a_2^2 & a_3^2 \\ a_1^3 & a_2^3 & a_3^3 \end{pmatrix} \begin{pmatrix} a_1^1 & a_1^2 & a_1^3 \\ a_2^1 & a_2^2 & a_2^3 \\ a_3^1 & a_3^2 & a_3^3 \end{pmatrix} = \begin{pmatrix} 1 & 0 & 0 \\ 0 & 1 & 0 \\ 0 & 0 & 1 \end{pmatrix} \tag{1.38}$$

and

$$\begin{pmatrix} a_1^1 & a_1^2 & a_1^3 \\ a_2^1 & a_2^2 & a_2^3 \\ a_3^1 & a_3^2 & a_3^3 \end{pmatrix} \begin{pmatrix} a_1^1 & a_2^1 & a_3^1 \\ a_1^2 & a_2^2 & a_3^2 \\ a_1^3 & a_2^3 & a_3^3 \end{pmatrix} = \begin{pmatrix} 1 & 0 & 0 \\ 0 & 1 & 0 \\ 0 & 0 & 1 \end{pmatrix} \tag{1.39}$$

Expanding each term of the indicated matrix multiply and equating to its corresponding element of the unit matrix, we have the following useful relationships for future reference. First from Eq. (1.38),

$$\begin{aligned}
a_1^1 a_1^1 + a_2^1 a_2^1 + a_3^1 a_3^1 &= 1 \\
a_1^1 a_1^2 + a_2^1 a_2^2 + a_3^1 a_3^2 &= 0 \\
a_1^1 a_1^3 + a_2^1 a_2^3 + a_3^1 a_3^3 &= 0 \\
a_1^2 a_1^1 + a_2^2 a_2^1 + a_3^2 a_3^1 &= 0 \\
a_1^2 a_1^2 + a_2^2 a_2^2 + a_3^2 a_3^2 &= 1 \\
a_1^2 a_1^3 + a_2^2 a_2^3 + a_3^2 a_3^3 &= 0 \\
a_1^3 a_1^1 + a_2^3 a_2^1 + a_3^3 a_3^1 &= 0 \\
a_1^3 a_1^2 + a_2^3 a_2^2 + a_3^3 a_3^2 &= 0 \\
a_1^3 a_1^3 + a_2^3 a_2^3 + a_3^3 a_3^3 &= 1
\end{aligned} \tag{1.40}$$

and then again from Eq. (1.39),

$$\begin{aligned}
a_1^1 a_1^1 + a_1^2 a_1^2 + a_1^3 a_1^3 &= 1 \\
a_1^1 a_2^1 + a_1^2 a_2^2 + a_1^3 a_2^3 &= 0 \\
a_1^1 a_3^1 + a_1^2 a_3^2 + a_1^3 a_3^3 &= 0
\end{aligned}$$

$$a_2^1 a_1^1 + a_2^2 a_1^2 + a_2^3 a_1^3 = 0$$
$$a_2^1 a_2^1 + a_2^2 a_2^2 + a_2^3 a_2^3 = 1$$
$$a_1^1 a_3^1 + a_2^2 a_3^2 + a_2^3 a_3^3 = 0$$

$$a_3^1 a_1^1 + a_3^2 a_1^2 + a_3^3 a_1^3 = 0$$
$$a_3^1 a_2^1 + a_3^2 a_2^2 + a_3^3 a_2^3 = 0$$
$$a_3^1 a_3^1 + a_3^2 a_3^2 + a_3^3 a_3^3 = 1 \qquad (1.41)$$

1.2.2 Determinant Equal to 1

Another useful property of the elements, a_i^α, of the a matrix may be realized by examining its determinant. From matrix algebra, the definition of the determinant of a matrix (Ref. 6, Sec. 4-1) may be written in the following form:

$$\det(a) = a_i^\alpha A_\alpha^i \qquad (1.42)$$

where A_α^i is the determinant of the cofactor of the a_i^α element of the a matrix and here i is not summed. The determinant is formed by sums on any row or column of the a matrix, i.e., Eq. (1.42) becomes

$$\det(a) = a_1^\alpha A_\alpha^1 = a_2^\alpha A_\alpha^2 = a_3^\alpha A_\alpha^3$$

or

$$\det(a) = a_\alpha^1 A_1^\alpha = a_\alpha^2 A_2^\alpha = a_\alpha^3 A_3^\alpha \qquad (1.43)$$

The elements of the A matrix are simply the corresponding minors of the a matrix and are given by the equation

$$A_i^j = (-1)^{i+j} M_{ij} \qquad (1.44)$$

where M_{ij} is the i-jth minor of the a matrix. We can construct a matrix from Eq. (1.42) by writing

$$(a)(A)^T = \det(a)(I) \qquad (1.45)$$

where I is the unit matrix, and remembering Eq. (1.37),

$$\det((a)(a)^T) = \det(I) = 1$$

and

$$\det(a)(\det(a)^T) = 1 \qquad (1.46)$$

but from basic matrix algebra, $\det(a) = \det((a)^T)$, then from Eq. (1.46), we must have

$$\det(a) = 1 \qquad (1.47)$$

1.2.3 Each Element Equal to Its Minor

Using Eq. (1.46) in Eq. (1.45), we can write

$$(a)(A^T) = 1(I) = (I)$$

and multiplying through by the inverse of a, which is simply its transpose, we find that $(A^T) = (a^T)$ or, in component form,

$$a_i^\alpha = A_{i\alpha} = (-1)^{i+\alpha} M_{i\alpha} \tag{1.48}$$

This states that each element of the a matrix is equal to its cofactor or its minor (Ref. 6, p. 75). For future reference, we can write out the equations of (1.48), giving us the following equalities:

$$\begin{aligned}
a_1^1 &= a_2^2 a_3^3 - a_3^2 a_2^3 \\
a_1^2 &= a_3^1 a_2^3 - a_2^1 a_3^3 \\
a_1^3 &= a_2^1 a_3^2 - a_3^1 a_2^2 \\
a_2^1 &= a_3^2 a_1^3 - a_1^2 a_3^3 \\
a_2^2 &= a_1^1 a_3^3 - a_3^1 a_1^3 \\
a_2^3 &= a_3^1 a_1^2 - a_1^1 a_3^2 \\
a_3^1 &= a_1^2 a_2^3 - a_2^2 a_1^3 \\
a_3^2 &= a_2^1 a_1^3 - a_1^1 a_2^3 \\
a_3^3 &= a_1^1 a_2^2 - a_2^1 a_1^2
\end{aligned} \tag{1.49}$$

1.2.4 Repeated Orthogonal Coordinate Transformations

Consider the following two coordinate systems transformations where only rotations are of interest:

$$x_i = a_i^\alpha \bar{x}_\alpha \tag{1.50}$$

and

$$\bar{x}_\alpha = b_\alpha^\gamma \bar{\bar{x}}_\gamma \tag{1.51}$$

Substituting Eq. (1.51) into Eq. (1.50),

$$x_i = a_i^\alpha b_\alpha^\gamma \bar{\bar{x}}_\gamma \tag{1.52}$$

The double-barred coordinates are now transformed to the unbarred coordinates by

$$x_i = c_i^\gamma \bar{\bar{x}}_\gamma \tag{1.53}$$

where the elements of new transformation matrix c are defined by

$$c_i^\gamma = a_i^\alpha b_\alpha^\gamma \tag{1.54}$$

when equating Eqs. (1.52) and (1.53). Equation (1.54) defines the matrix times a matrix operation, $(c) = (a)\,(b)$, in matrix notation. Going one step further, imagine a third coordinate system referenced to the double-barred system,

$$\bar{\bar{x}}_\gamma = d_\gamma^\rho y_\rho \tag{1.55}$$

and placing this into Eq. (1.52) such that

$$x_i = a_i^\alpha b_\alpha^\gamma d_\gamma^\rho y_\rho \tag{1.56}$$

and so forth. The components of an e matrix could be defined

$$e_i^\rho = a_i^\alpha b_\alpha^\gamma d_\gamma^\rho \tag{1.57}$$

Then Eq. (1.56) can be written

$$x_i = e_i^\rho y_\rho \tag{1.58}$$

which has precisely the same form as Eq. (1.50); hence after repeated coordinate transformations, the form of the equation remains the same. These are the characteristics of a Cartesian tensor equation. Notice the transformation Eqs. (1.52) and (1.53) and how these equation forms correspond to the general tensor form given in Eq. (1.23). Also note that after repeated matrix multiplications of orthogonal matrices, a single orthogonal matrix results, which defines the transformation.

Equation (1.54) develops each element of the c matrix as a function of the elements of the a and b matrices. Likewise, since these matrices are orthogonal, the following component equations for matrix a and matrix b can be written:

$$a_i^\alpha = c_{i\gamma} b^{\alpha\gamma} \tag{1.59}$$

which is equivalent to $(a) = (c)(b)^T$ in matrix notation. Matrix b could be written

$$b_\alpha^\gamma = a_{\beta\alpha} c^{\beta\gamma} \tag{1.60}$$

which is the same as $(b) = (a)^T(c)$ in matrix notation.

1.2.5 Computing Radius Vector Measurements

We can write Eq. (1.1) in what is defined as the *metric form* as follows:

$$r^2 = x_1^2 + x_2^2 + x_3^2 = g_{\alpha\beta} x^\alpha x^\beta \tag{1.61}$$

where the generalized equation of the right side characterizes what is called *Riemannian space* (Ref. 2, Introduction, Sec. 2). Notice that for Cartesian coordinates, $g_{\alpha\beta} = 1$ when $\alpha = \beta$ and $g_{\alpha\beta} = 0$ for all $\alpha \neq \beta$, a special case of the more general form of space representation. The Eucleadian representation of Eq. (1.1) will be used throughout the text with references to the more general form that is fundamental to geometry of space and time.

In many physical applications of engineering analysis, actual measurement of the distances x_1, x_2, and x_3 may not be possible, however, the vector distance $\boldsymbol{R}$ is known and the relative distance $\bar{\boldsymbol{r}}$ is readily measured. Therefore, to compute r^2 from Eq. (1.25), substitute x_i from Eq. (1.14), and we have

$$r^2 = g^{\alpha\beta}\left(a_\alpha^\gamma \bar{x}_\gamma + X_\alpha\right)\left(a_\beta^\rho \bar{x}_\rho + X_\beta\right) \tag{1.62}$$

By expanding the right-hand side of this equation, some interesting and useful relationships can be derived, starting with

$$r^2 = g^{\alpha\beta}\left(a_\alpha^\gamma a_\beta^\rho \bar{x}_\gamma \bar{x}_\rho + \left(a_\alpha^\gamma \bar{x}_\gamma X_\beta + a_\beta^\rho \bar{x}_\rho X_\alpha\right) + X_\alpha X_\beta\right) \tag{1.63}$$

Remembering from the preceding discussion that $\alpha = \beta$, we can carry out the indicated summations in the first term of Eq. (1.63) and use all of the relationships given in Eq. (1.40). We then have

$$g^{\alpha\alpha} a_\alpha^\gamma a_\alpha^\rho \bar{x}_\gamma \bar{x}_\rho = g^{\alpha\alpha} \bar{x}_\alpha \bar{x}_\alpha \tag{1.64}$$

The $g^{\alpha\alpha}$, although each equal to 1, are retained to indicate the summations on α. Equation (1.64) demonstrates the important relationship that is another defining characteristic of Cartesian tensors. Equation (1.64) forms the scalar product or the dot product and shows that *the scalar product is invariant during coordinate transformations* (Ref. 3, p. 12). Here, the length of a vector is again shown to be invariant as seen from one Cartesian coordinate frame to another, as used in Eq. (1.8).

Equation (1.63) then can be further reduced to

$$r^2 = g^{\alpha\beta}\left(\bar{x}_\alpha \bar{x}_\beta + \left(a_\alpha^\rho \bar{x}_\rho X_\beta + a_\beta^\gamma \bar{x}_\gamma X_\alpha\right) + X_\alpha X_\beta\right) \tag{1.65}$$

Again when $\alpha = \beta$, this equation is recognized as the Law of Cosines, and from Fig. 1.1 Eq. (1.65) becomes

$$r^2 = \bar{r}^2 + R^2 + 2\bar{r}R\cos\theta \tag{1.66}$$

where θ is the angle between the $\bar{\boldsymbol{r}}$ and $\boldsymbol{R}$ vectors. Again notice that

$$\bar{r}R\cos\theta = g^{\alpha\beta}\left(a_\alpha^\rho \bar{x}_\rho\right)X_\beta \tag{1.67}$$

is the definition for the dot product in vector algebra. In the dot products formed by $g^{\alpha\beta}\bar{x}_\alpha\bar{x}_\beta$ and $g^{\alpha\beta}X_\alpha X_\beta$, the angles between the vectors and themselves are zero, hence $\cos\theta = 1$.

A useful approximation for the radius vector magnitude $\boldsymbol{r}$, when the radius vector $\boldsymbol{R}$ is *very very large* and the vectors $\boldsymbol{r}$ and $\boldsymbol{R}$ become nearly parallel, is simply

$$r \cong R + \bar{r}\cos\theta \tag{1.68}$$

where $\cos\theta$ is given in the dot product of Eq. (1.67). This can also be visualized from vector geometry, and we can rewrite Eq. (1.68) as follows:

$$r \cong R + \frac{(\tilde{\boldsymbol{r}} \cdot \boldsymbol{R})}{R} \tag{1.69}$$

The inertial components of the $\tilde{\boldsymbol{r}}$ vector, from Eq. (1.62), are $\tilde{x}_i = a_i^{\alpha}\bar{x}_{\alpha}$.

1.2.6 Astrodynamics Application: Computational Star Navigation

The objective of this application is to describe a method that can be used to determine the attitude orientation of the space vehicle relative to the Earth-centered, inertial (ECI) reference frame. This method uses space vehicle-relative star observations to known stars that are in fixed directions in an ECI coordinate frame defined at a specified *epoch* or time.

We again refer to Fig. 1.1, where the center of the coordinate frame at O is the center of the Earth and the center of an inertial coordinate frame. As in Sec. 1.1.4, we can select the x_1 axis to point to the vernal equinox, which is defined using the line of intersection of the Earth's equatorial plane and the plane of the Earth's orbit around the sun. The plane of the Earth's orbit about the sun is called the *ecliptic plane*. The vernal equinox is a relatively stationary point where the path of the sun on the celestial sphere crosses the celestial equator when the sun is going northward. In other words, this is where the sun is on the first day of spring each year.

The Earth's rotational axis precesses in inertial space, and as a result of this motion the vernal equinox moves very slowly westward along the ecliptic plane. We can select a time or a common epoch and define the vernal equinox to be frozen or fixed at that epoch. The precessional motion of the Earth's axis is described very accurately by the precession transformation matrix (Ref. 5, p. B18),

$$\boldsymbol{r}'_t = (\mathcal{P})\boldsymbol{r}_{t_o} \tag{1.70}$$

where the $\mathcal{P}$ matrix, for example, transforms vectors in the reference ECI system, $\boldsymbol{r}_{t_o}$, at the given epoch, t_o, to vector components $\boldsymbol{r}'_t$ in the newly defined ECI frame at epoch t. The $\mathcal{P}$ matrix is a function of the time displacement from the reference epoch, i.e., a function of $(t - t_o)$.

The Earth's rotational axis also has small cycloid-like osculating motions about the precessing axis that are called *nutations*. These motions are described by the nutation transformation matrix,

$$\boldsymbol{r}_t = (\mathcal{N})\boldsymbol{r}'_t \tag{1.71}$$

where the $\mathcal{N}$ matrix accounts for the slight osculations from the precessing axis of the Earth at epoch t. The small nutation angles and methods to compute the nutation transformation matrix are given in Ref. 7. Using Eq. (1.52), the *nutation-precession transformation matrix* becomes

$$\boldsymbol{r}_t = (\mathcal{N})(\mathcal{P})\boldsymbol{r}_{t_o} \tag{1.72}$$

The elements of the $(\mathcal{N})(\mathcal{P})$ matrix are tabulated for each day of the year in Ref. 5 (pp. B34–B49). For an excellent detailed discussion of the angular motions of the Earth's axis due to precession and nutation, see Ref. 8 (Secs. 3.7.2–3.7.8).

Hence, we can define an ECI coordinate frame, at a given epoch, as the stationary frame to reference pointing directions in inertial space. However, when we refer to an ECI frame, it is necessary to attach a time-tag to that ECI frame for precise identification. For example, the ECI, true-of-date (TOD) frame is defined with its x axis pointing to the vernal equinox position at Greenwich midnight of the reference day. A second example, the ECI, J2004.5 frame, references the vernal equinox position on midnight of July 2.125, 2004 (Ref. 5, p. H1), or precisely the middle of the year 2004.

The bright stars and astronomical objects are located in ECI reference frames at specified epochs by two angles, the right ascension and the declination. The location of many of the bright stars are known and published each year in *The Astronomical Almanac* (Ref. 5, pp. H2–H31) using ECI reference frames at specified epochs.

A second coordinate frame can be defined as in Sec. 1.1.4, only this time let its center at O' be the center of mass of a space vehicle that is free to rotate in space relative to the inertial frame. Again imagine a space traveler seated in the space vehicle such that looking forward is the $\bar{x}_1$ axis, his extented right hand points in the $\bar{x}_2$-axis direction, and the $\bar{x}_3$ axis is pointing out the bottom of the space traveler's seat. This coordinate reference frame is called the *body axis frame* and is depicted at O' in Fig. 1.3.

We now define the body-axis-to-inertial-coordinate transformation matrix, $\boldsymbol{x} = (b)\bar{\boldsymbol{x}}$, which transforms vectors measured in the body axis frame into vectors in the inertial frame.

The space vehicle's body axis attitude can be held stationary relative to the inertial frame. This space vehicle controlled attitude mode is referred to specifically as *inertial attitude hold*. With the space vehicle in inertial-attitude-hold mode, the relative positions of stars can be accurately determined with an optical star tracker that can provide angular measurements to several stars even at the same time. The angular measurements can be used to compute unit vectors, $\hat{\boldsymbol{x}}_i$, to these stars. The star tracker optical sensor is mounted rigidly to the space vehicle body axis so that the unit vector measurements can be transformed from the star tracker frame to

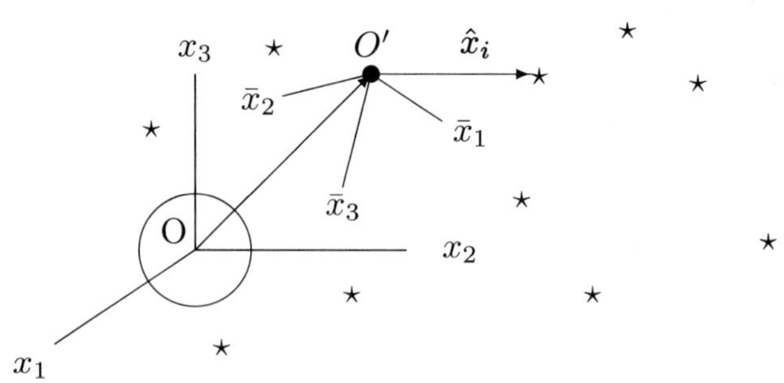

Fig. 1.3 Body axis to inertial coordinate.

the body axis frame by $\bar{\boldsymbol{x}} = (s)\hat{\boldsymbol{x}}$. Then analogous to Eq. (1.52), unit vectors to the star of opportunity in the inertial frame become

$$x_i = b_i^\alpha s_\alpha^\gamma \hat{x}_\gamma \tag{1.73}$$

or simply,

$$x_i = c_i^\gamma \hat{x}_\gamma \tag{1.74}$$

We know the unit vectors to the fixed positions of the stars on the celestial sphere, as defined in the ECI frame *of the selected epoch*. We can compute the unit vectors to those stars as seen from the star tracker and then compute the c matrix using the following analysis: first, place the known unit vectors to the selected stars $\boldsymbol{x}_i$ into a matrix, one unit vector per column, then place the measured unit vectors from the star tracker simularily into a matrix and factor out the c matrix yielding a matrix equation of the following form:

$$\begin{pmatrix} x_{1(1)} & x_{1(2)} & \cdots & x_{1(n)} \\ x_{2(1)} & x_{2(2)} & \cdots & x_{2(n)} \\ x_{3(1)} & x_{3(2)} & \cdots & x_{3(n)} \end{pmatrix} = \begin{pmatrix} c_1^1 & c_1^2 & c_1^3 \\ c_2^1 & c_2^2 & c_2^3 \\ c_3^1 & c_3^2 & c_3^3 \end{pmatrix} \begin{pmatrix} \hat{x}_{1(1)} & \hat{x}_{1(2)} & \cdots & \hat{x}_{1(n)} \\ \hat{x}_{2(1)} & \hat{x}_{2(2)} & \cdots & \hat{x}_{2(n)} \\ \hat{x}_{3(1)} & \hat{x}_{3(2)} & \cdots & \hat{x}_{3(n)} \end{pmatrix} \tag{1.75}$$

This equation can be written somewhat simpler in matrix form as

$$(X) = (c)(\hat{X}) \tag{1.76}$$

where the X matrix is not to be confused with the inertial vector to the center of mass of the space vehicle, but indeed represents the left-hand matrix of Eq. (1.75). It is possible to solve for the c matrix by applying the concepts of least-squares analysis (Ref. 8, pp. 676–681) to linear matrix algebra by first post-multiplying Eq. (1.76) by $(\hat{X})^T$. Notice from least-squares analysis that it will take at least three or more simultaneous star measurements for each body axis attitude matrix determination, i.e., $n \geq 3$, in Eq. (1.75). This operation will form a 3×3 square matrix on the right-hand side as follows:

$$(X)(\hat{X})^T = (c)((\hat{X})(\hat{X})^T) \tag{1.77}$$

and the c matrix is found by post-multiplying by the inverse of $(\hat{X})(\hat{X})^T$,

$$(c) = ((X)(\hat{X})^T)((\hat{X})(\hat{X})^T)^{-1} \tag{1.78}$$

Remembering Eq. (1.73), the body-axis-to-inertial transformation matrix becomes

$$(b) = (c)(s)^T \tag{1.79}$$

Thus by using an optical star tracker, the space vehicle's body axis coordinate frame can be precisely related to the stars and to the inertial coordinate frame, i.e., the ECI frame at the selected epoch t.

1.3 Euler Angles and the Transformation Matrix

1.3.1 Classical Euler Angle Set

When the nine dot products, the a_i^j or the direction cosines of the transformation, are in the form of Eq. (1.11), they are difficult to work with in the analysis of physical problems. Leonhard Euler, a Swiss mathematician and physicist (1707–1783), defined three angles of rotation that greatly simplify the definition of the a_i^j and can readily be visualized as follows:

1) Assume first that each of the three coordinate axes of two Cartesian coordinate frames are respectively coincident; then rotate one of the frames about the x_3 (or $\bar{x}_3'$) axis through the angle θ_1, as shown in Fig. 1.4.

2) Now rotate about the $\bar{x}_1'$ axis by the angle θ_2, as shown in Fig. 1.5.

3) Finally rotate about the new $\bar{x}_3$ axis by the angle θ_3, as shown in Fig. 1.6.

The angles θ_1, θ_2, and θ_3 define an Euler angle set describing a coordinate transformation. These angles are more practical to work with and easily lend themselves to definition in physical problems in contrast with the actual cosines of angles between each coordinate axis. Therefore, in order to apply Eqs. (1.14) and (1.15) to physical problems, it is necessary to determine the

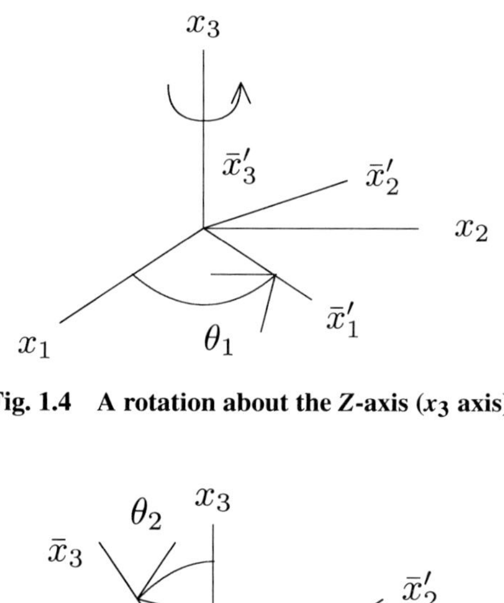

Fig. 1.4 A rotation about the *Z*-axis (x_3 axis).

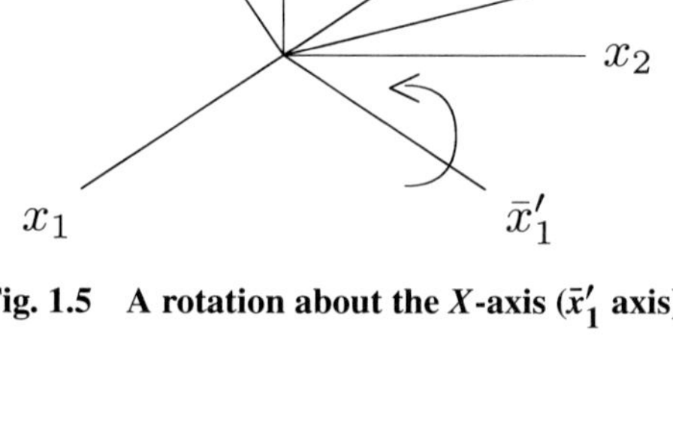

Fig. 1.5 A rotation about the *X*-axis ($\bar{x}_1'$ axis).

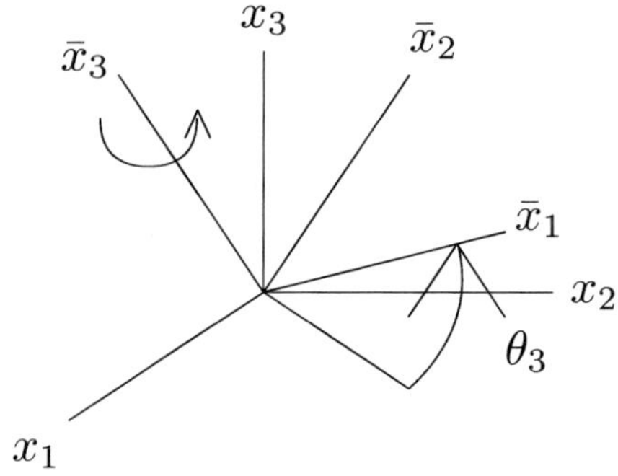

Fig. 1.6 A rotation about the Z-axis ($\bar{x}_3$ axis).

elements of the transformation matrix as functions of an Euler angle set such that $a_i^j(\theta_1, \theta_2, \theta_3)$.

The preceding particular sequence of rotations might be termed *the classical Euler angle set* and is presented in Appendix A.11. Many problems in astronomy and physics use this Euler set to define the transformation of vectors from an orbital coordinate frame to vectors in the inertial coordinate frame. The orbital plane in reference to the inertial frame can easily be visualized by realizing the following angular definitions:

1) $\theta_1 = \Omega$, the right ascension of the ascending node of the orbital plane,
2) $\theta_2 = i$, the inclination of the orbital plane, and
3) $\theta_3 = \omega$, the argument of the perigee of the orbit.

After close examination it will be shown that there are 12 possible Euler angle sequences using three rotations to define the transformation of vectors in the barred coordinate frame into vectors in the stationary frame. *All 12 Euler angle sets are equally as accurate in locating the barred coordinate frame.* However, a particular Euler set may lend itself more advantageous for application to a particular problem.

1.3.2 Single-Axis Rotations

If we examine each angular rotation singly, it is possible to generalize the solution for the a_i^j elements of the transformation matrix for any sequence of rotations including the classical Euler angle set described in Sec. 1.3.1 (Ref. 9, Sec. 48).

In an effort to standardize the use of transformation matrices, the matrix a is defined to transform vectors in the barred reference frame $(\bar{x}_1, \bar{x}_2, \bar{x}_3)$ into vectors in the stationary or inertial frame (x_1, x_2, x_3) as shown in Eq. (1.14). In this text, this transformation will be referred to as the *forward transformation*, whereas the inverse transformation, as in Eq. (1.15), will be referred to as the *reverse transformation*.

Referring only to the rotation part of Eq. (1.14), we can write in simple matrix form

$$\boldsymbol{x} = (a)\bar{\boldsymbol{x}} \tag{1.80}$$

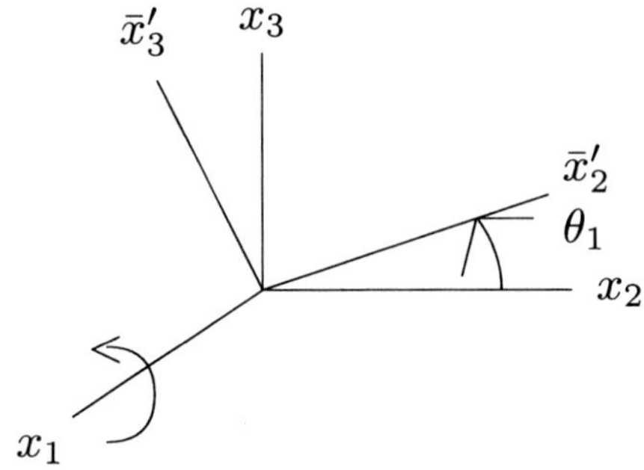

Fig. 1.7 The X-axis rotation.

Using the right-hand rule for positive rotations, the a matrix can be constructed by the following analysis. The first rotation, as shown in Fig. 1.7, is made about the stationary x_1 axis, referred to as an X-axis rotation, by the amount θ_1.

This single rotation about the x_1 axis (the X axis) results in the following transformation:

$$\begin{pmatrix} x_1 \\ x_2 \\ x_3 \end{pmatrix} = \begin{pmatrix} 1 & 0 & 0 \\ 0 & \cos\theta_1 & -\sin\theta_1 \\ 0 & \sin\theta_1 & \cos\theta_1 \end{pmatrix} \begin{pmatrix} \bar{x}'_1 \\ \bar{x}'_2 \\ \bar{x}'_3 \end{pmatrix} \tag{1.81}$$

or $\boldsymbol{x} = (X)\bar{\boldsymbol{x}}'$ in matrix form and here the X matrix represents the 3×3 matrix in Eq. (1.81). The X matrix here is not to be confused with the translation vector, for instance, in Eq. (1.14). A rotation of θ_2 about the $\bar{x}'_2$ axis (note: about an axis that has been rotated), referred to as a Y-axis rotation, is depicted in Fig. 1.8.

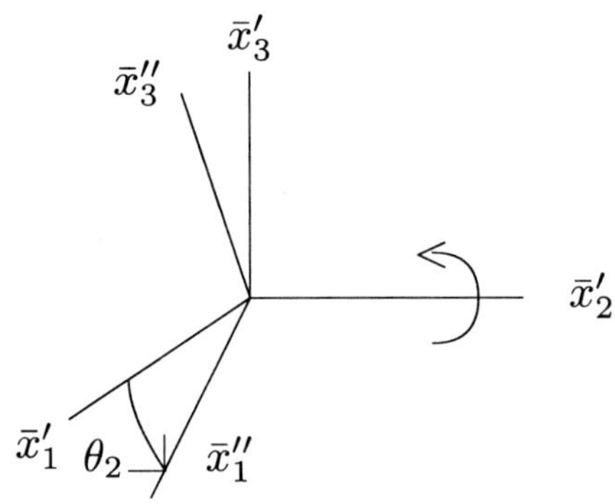

Fig. 1.8 The Y-axis rotation.

This rotation (the Y-axis rotation) yields the intermediate transformation matrix that transforms vectors from the $\bar{x}''$ frame into the $\bar{x}'$ by

$$\begin{pmatrix} \bar{x}'_1 \\ \bar{x}'_2 \\ \bar{x}'_3 \end{pmatrix} = \begin{pmatrix} \cos\theta_2 & 0 & \sin\theta_2 \\ 0 & 1 & 0 \\ -\sin\theta_2 & 0 & \cos\theta_2 \end{pmatrix} \begin{pmatrix} \bar{x}''_1 \\ \bar{x}''_2 \\ \bar{x}''_3 \end{pmatrix} \tag{1.82}$$

and in matrix form, $\bar{\boldsymbol{x}}' = (Y)\bar{\boldsymbol{x}}''$. Finally a rotation of θ_3 about the $\bar{x}''_3$ axis (note again: about an axis that has been rotated), referred to as a Z-axis rotation, similarly, is shown in Fig. 1.9.

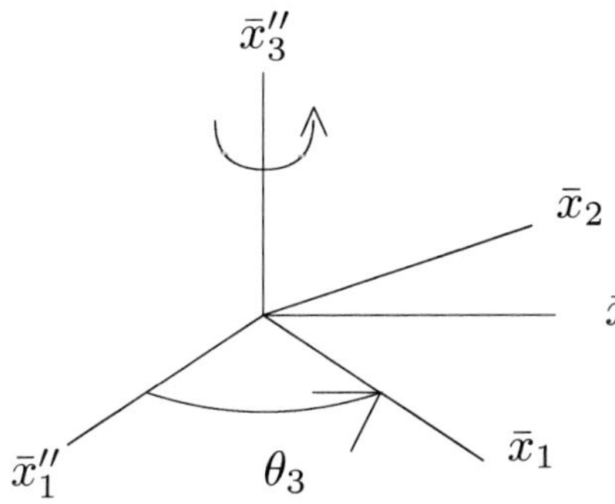

Fig. 1.9 The *Z*-axis rotation.

This final rotation (the Z-axis rotation) yields the intermediate transformation matrix:

$$\begin{pmatrix} \bar{x}''_1 \\ \bar{x}''_2 \\ \bar{x}''_3 \end{pmatrix} = \begin{pmatrix} \cos\theta_3 & -\sin\theta_3 & 0 \\ \sin\theta_3 & \cos\theta_3 & 0 \\ 0 & 0 & 1 \end{pmatrix} \begin{pmatrix} \bar{x}_1 \\ \bar{x}_2 \\ \bar{x}_3 \end{pmatrix} \tag{1.83}$$

and in matrix form, $\bar{\boldsymbol{x}}'' = (Z)\bar{\boldsymbol{x}}$. Using the three transformation equations,

$$\begin{aligned} \boldsymbol{x} &= (X)\bar{\boldsymbol{x}}' \\ \bar{\boldsymbol{x}}' &= (Y)\bar{\boldsymbol{x}}'' \\ \bar{\boldsymbol{x}}'' &= (Z)\bar{\boldsymbol{x}} \end{aligned} \tag{1.84}$$

and by back substitution,

$$\boldsymbol{x} = (X)(Y)(Z)\bar{\boldsymbol{x}} \tag{1.85}$$

Then from Eq. (1.80) we have the transformation matrix as a function of this Euler angle rotational sequence:

$$(a) = (X)(Y)(Z) \tag{1.86}$$

Computations for the elements of this transformation matrix from the indicated matrix multiplications of Eq. (1.86) yields

$$a_i^j = \begin{pmatrix} \cos\theta_2\cos\theta_3 & -\cos\theta_2\sin\theta_3 & \sin\theta_2 \\ \cos\theta_1\sin\theta_3 + \sin\theta_1\sin\theta_2\cos\theta_3 & \cos\theta_1\cos\theta_3 - \sin\theta_1\sin\theta_2\sin\theta_3 & -\sin\theta_1\cos\theta_2 \\ \sin\theta_1\sin\theta_3 - \cos\theta_1\sin\theta_2\cos\theta_3 & \sin\theta_1\cos\theta_3 + \cos\theta_1\sin\theta_2\sin\theta_3 & \cos\theta_1\cos\theta_2 \end{pmatrix} \tag{1.87}$$

This Euler rotation matrix defines the 1,2,3 Euler sequence of rotations, i.e., the $(X)(Y)(Z)$ sequence forming the *forward tranformation* matrix a, which is also given in Appendix A.1.

1.3.3 Three-Axis Rotations—The 12 Euler Rotation Matrices

Using the $(X)(Y)(Z)$ notation without the parentheses from Eq. (1.85), the following permutations of the three rotational sequences represent the 12 possible Euler angle sets:

$$\begin{array}{ccc} X\,Y\,Z & Y\,X\,Z & Z\,X\,Y \\ X\,Z\,Y & Y\,Z\,X & Z\,Y\,X \\ X\,Y\,X & Y\,X\,Y & Z\,X\,Z \\ X\,Z\,X & Y\,Z\,Y & Z\,Y\,Z \end{array} \tag{1.88}$$

Any repeated axis rotation, such as $X\,X\,Y$, simply reduces to the two-axis rotation $X\,Y$. The 12 Euler angle rotation sets of Eq. (1.88) can be used to define a transformation matrix. Conversely, there can be 12 three-axis Euler angle rotation sets that can be extracted from a transformation matrix.

The 12 Euler transformation matrices for three rotations are given in Appendix A. They are shown in the *forward* form, i.e., $\boldsymbol{x} = (a)\bar{\boldsymbol{x}}$. The *reverse* transformation matrix is found by simply the transpose of a, i.e., $\bar{\boldsymbol{x}} = (a)^T\boldsymbol{x}$, because $(a)^T = (a)^{-1}$, the orthogonal property of the transformation matrix.

In an effort to standardize the notation, it is suggested that the Euler angles occur in the same sequence as the rotations they define. Using this suggestion, the functional notation for Eq. (1.86) could be expressed as

$$(a) = (X)(Y)(Z) = X(\theta_x)Y(\theta_y)Z(\theta_z) = a(\theta_x, \theta_y, \theta_z) \tag{1.89}$$

and for the classical Euler angle set discussed in Sec. 1.3.1,

$$(a) = (Z)(X)(Z) = Z(\Omega)X(i)Z(\omega) = a(\Omega, i, \omega) \tag{1.90}$$

It is interesting to note that a negative rotation in the single-axis rotation matrices of Eqs. (1.81), (1.82), and (1.83) will result in the formation of the transpose of that matrix. However, the transpose of the transformation matrix a of Eq. (1.86) can be formed by reversing the order of multiplication and transposing each of the intermediate axis rotation matrices, i.e.,

$$(a)^T = ((X)(Y)(Z))^T = (Z)^T(Y)^T(X)^T \tag{1.91}$$

Using the notation in Eqs. (1.89) and (1.90), the Eq. (1.91) could be written

$$a^T(\theta_x, \theta_y, \theta_z) = a(-\theta_z, -\theta_y, -\theta_x) \tag{1.92}$$

However, to avoid confusion, it is recommended that the *forward* transformation be computed and simply transposed to yield the inverse transformation matrix.

1.3.4 Extracting Euler Angles

By examination of the elements of a transformation matrix and knowing the form, i.e., when three Euler rotations are being used in the analysis, any rotational sequence of Euler angles can be extracted from the matrix. However, inherent to each Euler angle set, there exists a discontinuous angular situation causing the angles to be ill defined. An example of this can be visualized by using Fig. 1.6. As the angle β_2 (the orbital inclination relative to the inertial plane) approaches zero, the angles β_1 and β_3 become undefined. These discontinuous situations may occur at 0 deg or at 90 deg depending on the Euler angle set being used in the analysis. This situation exists when the analyst is attemping to extract a three Euler angle sequence when, in this example, the transformation matrix is the result of a single-axis rotation. Being aware of these limitations and using care in interpreting the results, we can extract the Euler angle sets from a transformation matrix. For instance, if the *XYZ* Euler angle sequence is used from Eq. (1.87), the Euler angles are given by

$$\theta_1 = \tan^{-1}\left(\frac{a_2^3}{a_3^3}\right)$$

$$\theta_2 = \tan^{-1}\left(\frac{a_1^3}{\sqrt{1 - a_1^3}}\right)$$

and

$$\theta_3 = \tan^{-1}\left(\frac{a_1^2}{a_1^1}\right) \tag{1.93}$$

Notice that the arctangent function is used so that the expressions will be valid even when possible discontinuous angular situations occur. The arctangent function is also recommended to capture the full angular range of the Euler angles, i.e., from 0 to 360 deg. This can be accomplished by selecting the software arctangent function that yields the angular range, 0 to $\pm\pi$. Then, if the angle occurs between 0 and $-\pi$, simply add 2π to its value and the Euler angle range between 0 and 2π is achieved. Similiar equations can be written for each of the 12 three-axis Euler rotational sequences. Each of these Euler angle sets are presented in Appendix A.

1.3.5 Four-Axis Rotations/Multi-Axis Rotations

Transformation matrices may be formed by a single-axis rotation or by multi-axis rotations:

$$(a) = R_1 R_2 R_3 \dots R_n \tag{1.94}$$

where $R_1 R_2 R_3 \dots R_n$ form the products of single-axis rotation matrices describing the physical problem of interest. Equation (1.87) describes a typical Euler angle representation where $n = 3$ and Eqs. (1.81) through (1.83) are single-axis rotations where $n = 1$.

Extracting or solving for more than three Euler angles from a transformation matrix is difficult and ambiguous because of the transcendental nature of angular measurements. These procedures are problem specific, and usually some of the Euler angles, i.e., in excess of three, must be constrained to constant values in order to determine a remaining set of three angles. Other methods of angular representation can be more desirable, such as the use of quaternions, which will be discussed in Sec. 1.4.3.

1.3.6 Aerospace Application: Four-Axis Rotation Matrices Used to Determine Space Vehicle Attitudes

Example 1.1

A four-axis rotation system is used in the mechanical inertial measurement units[10] (IMU) to determine the attitude of a space vehicle. The development of the four-axis gimbal system was necessary to avoid the three-axis discontinuity (gimbal lock condition) when the system movement passes through the 0 or 90 deg angular singular point. These are the same mathematical singular points as described in Sec. 1.3.4 when extracting the Euler angles from a transformation matrix. A four-axis IMU system is shown in Fig. 1.10.

The rapidly spinning Z and X-Y gyroscopes create large angular momentum vectors or *spin vectors* pointing along the three coordinate axes of the platform. The angular momentum vectors tend to remain stationary and, hence, point in fixed directions in inertial space. The angular momentum of a spinning rigid body will be discussed in detail in Secs. 3.1.5 and 3.2.3 of the text. The gyroscopes are rigidly attached to the platform, which is called the *stable member*. When rotational motion occurs, the gyroscopic forces act to stabilize the platform, and the stable member will remain in a nearly fixed orientation in inertial space. The outer roll gimbal is mounted to the space vehicle's frame but is free to rotate about the outer roll gimbal resolvers. When the attitude of the space vehicle changes, all of the gimbal axes are free to rotate, while the stable platform remains fixed in inertial space. These angles are measured by the gimbal resolvers, shown in Fig. 1.10 by the black dots, and indicate the orientation of the platfrom (or the stable member) relative to the mounting base, which is called the *navigation base*.

If vectors in the stable member coordinate frame are $\mathbf{y}$ and vectors in the navigation base frame are $\bar{\mathbf{y}}$, the transformation matrix is

$$\mathbf{y} = (G)\bar{\mathbf{y}} \tag{1.95}$$

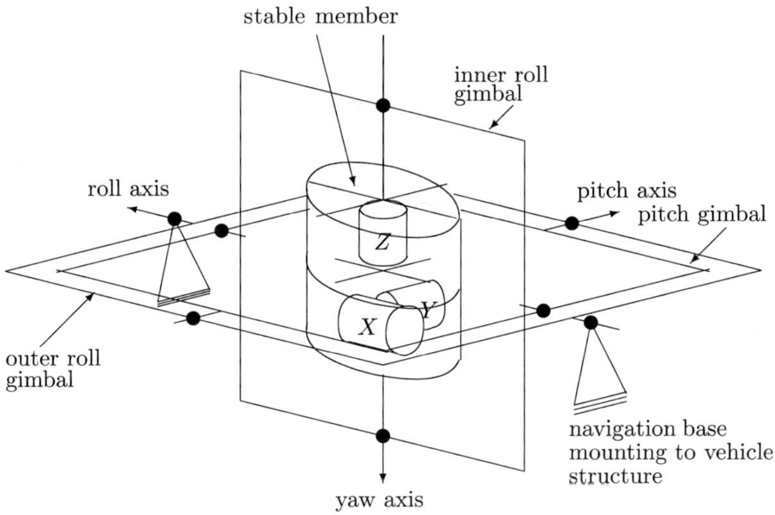

Fig. 1.10 The IMU with four gimbal axes.

A typical IMU matrix G is computed from the measurements of the angles read from the gimbal resolvers around the yaw, roll, pitch, and outer-roll axes, as shown in Fig. 1.10. Using these angles and single-axis rotations as described in Sec. 1.3.2, we have

$$(G) = Z(\text{yaw})X(\text{roll})Y(\text{pitch})X(\text{outer-roll}) \tag{1.96}$$

This models the four-axis Euler transformation matrix defined by the preceding single-axis gimbal rotational sequence.

To determine the actual space vehicle attitude relative to the inertial frame, the coordinate system diagram shown in Fig. 1.11 clarifies the matrix procedures. The

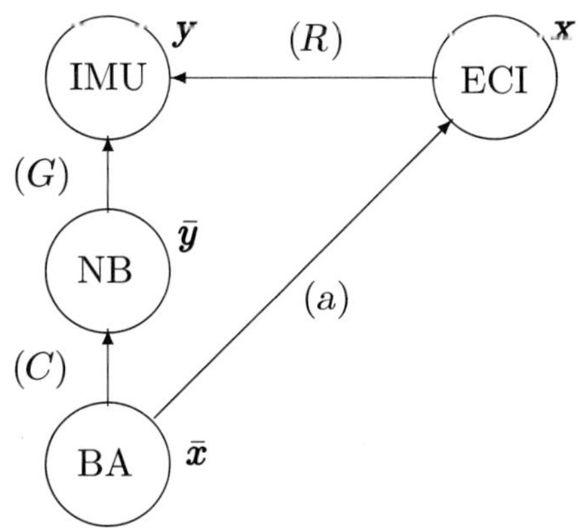

Fig. 1.11 The vehicle attitude coordinate transformations.

transformation matrix R is referred to as the *refs mat*, which transforms vectors from the inertial frame into vectors in the IMU frame:

$$\boldsymbol{y} = (R)\boldsymbol{x} \tag{1.97}$$

The IMU gyro-stabilized platform tends to precess slowly in time, causing the IMU to drift away from the original refs mat orientation at time zero. For the purpose of this discussion, the refs mat will be considered a constant transformation matrix, which is ideally fixed relative to inertial space. When the IMU platform is released or "un-caged," the gimbal axes are free to move relative to the navigation base. The transformation

$$\bar{\boldsymbol{x}} = (C)\bar{\boldsymbol{y}} \tag{1.98}$$

is a constant matrix that is defined by the mounting structure of the IMU to the space vehicle's structure. The space vehicle's body-axis-to-inertial-coordinate transformation matrix, which will give us the attitude of the space vehicle, is

$$\boldsymbol{x} = (a)\bar{\boldsymbol{x}} \tag{1.99}$$

To compute the a matrix as a function of the four gimbal angles, we first rewrite Eq. (1.97),

$$\boldsymbol{x} = (R)^T\boldsymbol{y} \tag{1.100}$$

and using Eqs. (1.95) and (1.98), we have

$$\boldsymbol{x} = ((R)^T(G)(C)^T)\bar{\boldsymbol{x}} \tag{1.101}$$

Therefore,

$$(a) = (R)^T(G)(C)^T \tag{1.102}$$

by comparison with Eq. (1.99). The a matrix describes the body axis orientation relative to the inertial frame and hence defines the space vehicle's attitude.

In the late 1980s, the ring laser gyroscopes began to replace the electromechanical stable platforms. The ring laser gyroscope is able to measure inertial rotation rates by detecting the wavelength differences in the light from two counter-rotating laser beams when motion occurs. The analysis to determine space vehicle attitude from the ring laser IMU is different from the preceding analysis, but very accurate angular displacements can be accumulated from the initial orientation of the laser unit.

Example 1.2

The angular orientation of the local-vertical, local-horizontal coordinate frame (the LVLH frame) as discussed in Sec. 1.1.4 can also be defined using Euler angle rotations. By examining the orientation of the orbital plane referenced to the Earth-centered, inertial frame as shown in Fig. 1.6 (the classical Euler angle set), we simply need one more orbital parameter, namely the orbit longitude of the space vehicle in its orbit. The orbit longitude ℓ is the angular position of the space vehicle measured in the orbital plane from the ascending node Ω of the orbit. With these three orbital parameters, Ω, i, and ℓ, the transformation matrix a of Eq. (1.25) can be constructed with the following single-axis rotation matrices using the same notation as in Eq. (1.90):

$$(a) = Z(\Omega)X(i)Z\left(\ell + \frac{\pi}{2}\right)X\left(\frac{3\pi}{2}\right) \tag{1.103}$$

Since the last rotation $X\left(\frac{3\pi}{2}\right)$ is through a known angle, i.e., 270 deg, the three Euler rotation angles can be extracted without ambiguity from the following three-axis intermediate matrix using the LVLH to ECI transformation matrix:

$$(a') = (a)\left(X\left(\frac{3\pi}{2}\right)\right)^T \tag{1.104}$$

It will be shown later in the text that the angle Ω, the right ascension of the orbit's ascending node, and the angle i, the orbital inclination, are constants for the space vehicle's *Keplerian orbit*. Also for space vehicles in nearly circular orbits, the longitude ℓ varies at a constant rate, which is specifically referred to as the *orbital rate*, namely, 2π divided by the orbital period. Hence, once the space vehicle's $\bar{x}_1$ body axis is oriented in the velocity direction and the $\bar{x}_2$ body axis is pointing perpendicular out of the orbital plane, the body axis can be held in the LVLH attitude by simply pitching down at the orbital rate. This body axis controlled attitude hold mode is referred to as the *LVLH-attitude-hold* mode.

1.4 Euler's Theorem and Four Parameter Methods

1.4.1 Euler's Coordinate Rotation Theorem

Although the Euler angles, as described in Sec. 1.3, are widely used in the solution of many problems in mechanics and engineering, the mathematical singular condition that exists in each Euler angle set sometimes presents problems. Leonhard Euler developed the following concepts for coordinate axes rotations in 1776, late in his life at age 69. This concept of axis rotation using four angular parameters, instead of the three Euler angles, is one of Euler's most outstanding contributions to the understanding of rotational dynamics. Stated simply, Euler's Theorem (Ref. 11, pp. 2, 3) on coordinate axis rotation is as follows: *Any instantaneous rotation of a coordinate frame may be represented by a single rotation about a single line through the origin of the coordinate frame.*

Consider Fig. 1.12, where one coordinate system has been rotated through the classical Euler angle set, θ_1, θ_2, and θ_3, while the coordinate centers remain coincidental at the *instantaneous center of rotation*. Euler identified a single line, which is stationary in both coordinate frames, about which the original coordinate frame is rotated to define the new coordinate frame. This line is called the *instantaneous axis of rotation* and is sometimes referred to as the *eigenaxis* of the coordinate transformation. Figure 1.12 shows the single rotation of ω about the single axis of rotation, the vector $\boldsymbol{R}$.

The eigenaxis is defined in space by the direction angles, α, β, and γ, measured from each of the coordinate axes x_1, x_2, and x_3, respectively. When the coordinate

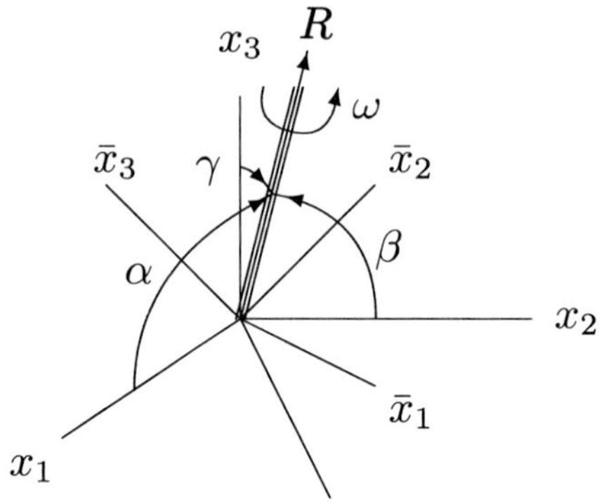

Fig. 1.12 The eigen axis rotation.

frame is rotated about the vector $\boldsymbol{R}$, the eigenaxis, through the angle ω, each coordinate axis, x_1, x_2, and x_3, will be rotated as the elements of cones, each through the same cone angle ω. Euler derived (Ref. 11, pp. 8, 9) the elements of the transformation matrix a_i^j as functions of the direction angles α, β, γ and the rotation angle ω as follows:

$$
\begin{aligned}
a_1^1 &= 1.0 - 2\sin^2(1/2)\omega \sin^2\alpha \\
a_2^1 &= 2\cos\alpha\cos\beta\sin^2(1/2)\omega + 2\cos(1/2)\omega\sin(1/2)\omega\cos\gamma \\
a_3^1 &= 2\cos\alpha\cos\gamma\sin^2(1/2)\omega - 2\cos(1/2)\omega\sin(1/2)\omega\cos\beta \\
a_1^2 &= 2\cos\alpha\cos\beta\sin^2(1/2)\omega - 2\cos\gamma\sin(1/2)\omega\cos(1/2)\omega \\
a_2^2 &= 1.0 - 2\sin^2(1/2)\omega\sin^2\beta \\
a_3^2 &= 2\cos\beta\cos\gamma\sin^2(1/2)\omega + 2\cos\alpha\sin(1/2)\omega\cos(1/2)\omega \\
a_1^3 &= 2\cos\alpha\cos\gamma\sin^2(1/2)\omega + 2\cos\beta\sin(1/2)\omega\cos(1/2)\omega \\
a_2^3 &= 2\cos\beta\cos\gamma\sin^2(1/2)\omega - 2\sin(1/2)\omega\cos(1/2)\omega\cos\alpha \\
a_3^3 &= 1.0 - 2\sin^2(1/2)\omega\sin^2\gamma
\end{aligned}
\tag{1.105}
$$

1.4.2 Transformation Matrix as a Function of the Euler Four Symmetric Parameters

Equations (1.105) can be simplified by using the parameters suggested by Euler in 1776 (Ref. 11, p. 8). The Euler four symmetric parameters are

$$
\begin{aligned}
\xi &= e_1 = \cos\alpha\sin(1/2)\omega \\
\eta &= e_2 = \cos\beta\sin(1/2)\omega
\end{aligned}
$$

$$\zeta = e_3 = \cos\gamma \sin(1/2)\omega$$
$$\chi = e_4 = \cos(1/2)\omega \tag{1.106}$$

Note that

$$e_1^2 + e_2^2 + e_3^2 + e_4^2 = 1 \tag{1.107}$$

since $\cos^2\alpha + \cos^2\beta + \cos^2\gamma = 1$. Using the equations of (1.106) and substituting them into Eqs. (1.105), we have the transformation matrix as a function of the Euler four symmetric parameters:

$$(a) = \begin{pmatrix} (e_1^2 - e_2^2 - e_3^2 + e_4^2) & 2(e_1e_2 - e_3e_4) & 2(e_1e_3 + e_2e_4) \\ 2(e_1e_2 + e_3e_4) & (-e_1^2 + e_2^2 - e_3^2 + e_4^2) & 2(e_2e_3 - e_1e_4) \\ 2(e_1e_3 - e_2e_4) & 2(e_2e_3 + e_1e_4) & (-e_1^2 - e_2^2 + e_3^2 + e_4^2) \end{pmatrix} \tag{1.108}$$

1.4.3 Hamilton Quaternion

Sir William Rowan Hamilton, an English mathematician, defined the quaternion and developed an entire algebra for operations using quaternions, as described in detail in his four volume treatise.[12] This work was done in about 1843 and was in some way a continuation of the concepts developed by Leonhard Euler. Felix Klein developed the Cayley-Klein parameters in 1897 (Ref. 13, Secs. 4–5; Ref. 14, pp. 5–9). This four-parameter method was introduced into classical mechanics in connection with the treatment of spatial rotations in quantum mechanics. A discussion of Cayley-Klein parameters is beyond the scope of this text; however, they are mentioned here because of their early use and importance to rotational dynamics. The Hamilton quaternion and its applications to modern aerospace problems are the primary interests of this text.

The Hamilton quaternion is very similar to the Euler four symmetric parameters and has the following correspondence:

$e_1 = q_2$		$q_1 = \cos(1/2)\omega$
$e_2 = q_3$		$q_2 = \sin(1/2)\omega\cos\alpha$
$e_3 = q_4$	Hamilton quaternion	$q_3 = \sin(1/2)\omega\cos\beta$
$e_4 = q_1$		$q_4 = \sin(1/2)\omega\cos\gamma$ (1.109)

Notice that the Hamilton quaternion also forms a unit element set, as in Eq. (1.107). Hamilton chose to think of the quaternion as a scalar plus a vector in the following form:

$$Q = q_1 + q_2\boldsymbol{i} + q_3\boldsymbol{j} + q_4\boldsymbol{k} \tag{1.110}$$

Using Eqs. (1.109), the a matrix can be written as a function of the Hamilton quaternion:

$$(a) = \begin{pmatrix} (q_1^2 + q_2^2 - q_3^2 - q_4^2) & 2(q_2q_3 - q_1q_4) & 2(q_2q_4 + q_3q_1) \\ 2(q_2q_3 + q_1q_4) & (q_1^2 - q_2^2 + q_3^2 - q_4^2) & 2(q_3q_4 - q_1q_2) \\ 2(q_2q_4 - q_1q_3) & 2(q_3q_4 + q_1q_2) & (q_1^2 - q_2^2 - q_3^2 + q_4^2) \end{pmatrix} \tag{1.111}$$

Unlike the 12 Euler angle rotational sequences to describe the transformation matrix as shown by Eqs. (1.88), only two quaternions can be found to define the same coordinate transformation of Eq. (1.111). They are simply (q_1, q_2, q_3, q_4) and $(-q_1, -q_2, -q_3, -q_4)$. These two quaternions represent a positive rotation about the rotation axis pointing in one direction and an opposite rotation about the same line of rotation except pointing in the opposite direction. Both of these quaternions will satisfy Eq. (1.111).

The early works by Hamilton suggested the following notation: let $q_1 = s$ and $\boldsymbol{v} = (q_2, q_3, q_4)$, so that Eq. (1.111) could be expressed as

$$(a) = a(q_1, q_2, q_3, q_4) = a(s, \boldsymbol{v}) \tag{1.112}$$

Continuing from this discussion,

$$a(s, \boldsymbol{v}) = a(-s, -\boldsymbol{v}) \tag{1.113}$$

The transpose of the transformation matrix is formed simply by

$$a^T(s, \boldsymbol{v}) = a(-s, \boldsymbol{v}) = a(s, -\boldsymbol{v}) \tag{1.114}$$

1.4.4 Transformation Matrix as a Function of the Hamilton Quaternion

Each element of the a matrix as shown in Eq. (1.111) contains second-order terms; therefore, the matrix may be broken down into simpler terms or factored. The following methods of transformation matrix representations are useful in the analysis of certain problems.

Borrowing tensor notation, let $a_i^\alpha = Q_i^{\alpha\rho} q_\rho$. Here ρ is summed from 1 to 4, and when $i = 1$, we can write

$$a_1^\alpha = Q_1^{\alpha\rho} q_\rho = \begin{pmatrix} q_1 & q_2 & -q_3 & -q_4 \\ -q_4 & q_3 & q_2 & -q_1 \\ q_3 & q_4 & q_1 & q_2 \end{pmatrix} \begin{pmatrix} q_1 \\ q_2 \\ q_3 \\ q_4 \end{pmatrix}$$

when $i = 2$,

$$a_2^\alpha = Q_2^{\alpha\rho} q_\rho = \begin{pmatrix} q_4 & q_3 & q_2 & q_1 \\ q_1 & -q_2 & q_3 & -q_4 \\ -q_2 & -q_1 & q_4 & q_3 \end{pmatrix} \begin{pmatrix} q_1 \\ q_2 \\ q_3 \\ q_4 \end{pmatrix}$$

and when $i = 3$,

$$a_3^\alpha = Q_3^{\alpha\rho} q_\rho = \begin{pmatrix} -q_3 & q_4 & -q_1 & q_2 \\ q_2 & q_1 & q_4 & q_3 \\ q_1 & -q_2 & -q_3 & q_4 \end{pmatrix} \begin{pmatrix} q_1 \\ q_2 \\ q_3 \\ q_4 \end{pmatrix} \tag{1.115}$$

It might be helpful to imagine Q_k^{ij} in three dimensions where there exists three planes (here the kth subscript defines the kth plane) each containing a 3×4 matrix with i and j being the row and column in the matrix of the kth plane.

Another useful relationship, also using the summation convention, is $a_i^\alpha = S_i^\beta T_\beta^\alpha$, which has its matrix equivalent as follows:

$$a_i^\alpha = \begin{pmatrix} q_2 & q_1 & -q_4 & q_3 \\ q_3 & q_4 & q_1 & -q_2 \\ q_4 & -q_3 & q_2 & q_1 \end{pmatrix} \begin{pmatrix} q_2 & q_3 & q_4 \\ q_1 & -q_4 & q_3 \\ q_4 & q_1 & -q_2 \\ -q_3 & q_2 & q_1 \end{pmatrix} \tag{1.116}$$

The S and T matrices of Eq. (1.116) are interesting because

$$(S)(S)^T = (I) \tag{1.117}$$

and

$$(T)^T(T) = (I) \tag{1.118}$$

where I is the 3×3 unit matrix.

Although it is beyond the scope of this text to present a detailed discussion of quaternion algebra using Hamilton's quaternion notation, it is important here to mention some of the fundamental concepts of this great work. As suggested by Hamilton, we can start with the interesting relationships representing the transformation matrix by using partitions of a 4×4 matrix, here written in tensor form using calligraphic capital letters to represent the 4×4 matrices:

$$\mathcal{A}_i^j = \mathcal{S}_i^\alpha \mathcal{T}_\alpha^j \tag{1.119}$$

where all indices are summed from 1 to 4 and $\mathcal{A}_i^j$ is defined by

$$\mathcal{A}_i^j = \begin{pmatrix} 1 & 0 & 0 & 0 \\ 0 & a_1^1 & a_1^2 & a_1^3 \\ 0 & a_2^1 & a_2^2 & a_2^3 \\ 0 & a_3^1 & a_3^2 & a_3^3 \end{pmatrix} \tag{1.120}$$

This requires that

$$\mathcal{S}_i^\alpha \mathcal{T}_\alpha^j = \begin{pmatrix} q_1 & -q_2 & -q_3 & -q_4 \\ q_2 & q_1 & -q_4 & q_3 \\ q_3 & q_4 & q_1 & -q_2 \\ q_4 & -q_3 & q_2 & q_1 \end{pmatrix} \begin{pmatrix} q_1 & q_2 & q_3 & q_4 \\ -q_2 & q_1 & -q_4 & q_3 \\ -q_3 & q_4 & q_1 & -q_2 \\ -q_4 & -q_3 & q_2 & q_1 \end{pmatrix} \tag{1.121}$$

Using the S and T matrices from Eq. (1.116) in partition form, we have

$$\begin{pmatrix} 1 & 0 \\ 0 & (a) \end{pmatrix} = \begin{pmatrix} q_1 & -q_2 & -q_3 & -q_4 \\ & (S) & & \end{pmatrix} \begin{pmatrix} q_1 & \\ -q_2 & (T) \\ -q_3 & \\ -q_4 & \end{pmatrix} \tag{1.122}$$

The difficulty in working with this representation is that the matrices are 4×4 in size, and most physical problems of Cartesian analysis involve three component vectors and 3×3 matrices. However, Hamilton's quaternion algebra was developed using the three vector components in a four-vector form, i.e., the vectors are treated as quaternions with zero scalar parts and thus utilizing the 4×4 operations with the matrices of Eq. (1.121). The Hamilton *four-vector form* was used in early aerospace applications by Bean[14] in 1969, Ickes[15] in 1970, and Carroll[16] in 1975. Using Eq. (1.120) and as shown in Eq. (1.50), the coordinate transformation in four-vector form can be written

$$\begin{pmatrix} 0 \\ x_1 \\ x_2 \\ x_3 \end{pmatrix} = (\mathcal{A}) \begin{pmatrix} 0 \\ \bar{x}_1 \\ \bar{x}_2 \\ \bar{x}_3 \end{pmatrix} \tag{1.123}$$

It is also interesting to note that both the $\mathcal{S}$ and the $\mathcal{T}$ matrices are orthogonal, and therefore we can write

$$(\mathcal{S})^T(\mathcal{S}) = (\mathcal{S})(\mathcal{S})^T = (\mathcal{I}) \tag{1.124}$$

and

$$(\mathcal{T})^T(\mathcal{T}) = (\mathcal{T})(\mathcal{T})^T = (\mathcal{I}) \tag{1.125}$$

where the $\mathcal{I}$ matrix is the 4×4 unit matrix.

Based on these concepts, we now represent the quaternion as a scaler, where $s = q_1$, and a column vector $\boldsymbol{v} = (q_2, q_3, q_4)$ for its vector part, as alluded to earlier in the text in Eqs. (1.110) and (1.112). By examination of the skew-symmetric matrices of Eq. (1.121), we see that it can be partitioned as follows:

$$\mathcal{S}_i^\alpha \mathcal{T}_\alpha^j = \begin{pmatrix} s & -(\boldsymbol{v})^T \\ (\boldsymbol{v}) & (w_q) \end{pmatrix} \begin{pmatrix} s & (\boldsymbol{v})^T \\ -(\boldsymbol{v}) & (w_q) \end{pmatrix} \tag{1.126}$$

where the matrix partition w_q is defined by the 3×3 skew-symmetrix as follows:

$$(w_q) = s(I) + (w_{ss}) = s \begin{pmatrix} 1 & 0 & 0 \\ 0 & 1 & 0 \\ 0 & 0 & 1 \end{pmatrix} + \begin{pmatrix} 0 & -v_3 & v_2 \\ v_3 & 0 & -v_1 \\ -v_2 & v_1 & 0 \end{pmatrix}$$

$$= \begin{pmatrix} q_1 & -q_4 & q_3 \\ q_4 & q_1 & -q_2 \\ -q_3 & q_2 & q_1 \end{pmatrix} \tag{1.127}$$

As hinted to by all of the preceding qualities of the transformation matrix, Hamilton's quaternion algebra is based on the quaternion as an entity in itself and is defined analogous to Eqs. (1.114) as follows:

$$\tilde{q} = q_1 + q_2 i + q_3 j + q_4 k$$

and

$$\tilde{q}^* = q_1 - q_2 i - q_3 j - q_4 k \tag{1.128}$$

where i, j, and k *are no longer the conventional coordinate axis unit vectors*. The quaternion whose magnitude is equal to one is sometimes called a *versor*. The i, j, and k are defined as unit quantities, analogous to the imaginary numbers, which behave according to the following algebric operating rules:

$$\begin{aligned} i^2 &= -1 & \qquad ij &= -ji = k \\ j^2 &= -1 & jk &= -kj = i \\ k^2 &= -1 & ki &= -ik = j \end{aligned} \tag{1.129}$$

Using the algebraic distributive laws and the rules of Eq. (1.129), we have

$$\tilde{q}\tilde{q}^* = (q_1 + q_2 i + q_3 j + q_4 k)(q_1 - q_2 i - q_3 j - q_4 k) = 1$$

and

$$\tilde{q}^*\tilde{q} = 1 \tag{1.130}$$

This also defines the quaternion multiply operation. With these definitions, the transformation equation, as written in Eq. (1.123), is equivalent to

$$\begin{aligned} \tilde{x} &= \tilde{q}\tilde{\bar{x}}\tilde{q}^* \\ &= (q_1 + q_2 i + q_3 j + q_4 k)(0 + \bar{x}_1 i + \bar{x}_2 j + \bar{x}_3 k)(q_1 - q_2 i - q_2 j - q_4 k) \end{aligned} \tag{1.131}$$

when written using the quaternion algebra notation. The matrix structure in Eq. (1.123) is formed when carrying out the preceding indicated operations as follows: the 48 terms created by the operations in Eq. (1.131) are arranged with the scaler terms forming the first row of the matrix followed by the i, j, and kth rows forming the second, third, and fourth rows of the matrix, respectively.

1.4.5 Extracting the Hamilton Quaternion

When the coordinate transformation matrix $\boldsymbol{x} = (a)\bar{\boldsymbol{x}}$ is known, and the form of each element is given from Eq. (1.111), the quaternion can be determined using the analysis by Klumpp[17] from 1976 as follows. First examine the sums

$$
\begin{aligned}
a_1^1 + a_2^2 + a_3^3 &= 3q_1^2 - q_2^2 - q_3^2 - q_4^2 \\
a_1^1 - a_2^2 - a_3^3 &= -q_1^2 + 3q_2^2 - q_3^2 - q_4^2 \\
-a_1^1 + a_2^2 - a_3^3 &= -q_1^2 - q_2^2 + 3q_3^2 - q_4^2 \\
-a_1^1 - a_2^2 + a_3^3 &= -q_1^2 - q_2^2 - q_3^2 + q_4^2
\end{aligned} \tag{1.132}
$$

and remembering that the a_i^j are matrix elements, i.e., the superscripts are not exponents, and the q_i are components of the quaternion, i.e., the superscripts are exponents. Recall from Eq. (1.107) that $q_1^2 + q_2^2 + q_3^2 + q_4^2 = 1$, and solving each equation of (1.132) for the q_i, respectively, we have

$$
\begin{aligned}
q_1 &= \pm(1/2)\sqrt{a_1^1 + a_2^2 + a_3^3 + 1} \\
q_2 &= \pm(1/2)\sqrt{a_1^1 - a_2^2 - a_3^3 + 1} \\
q_3 &= \pm(1/2)\sqrt{-a_1^1 + a_2^2 - a_3^3 + 1} \\
q_4 &= \pm(1/2)\sqrt{-a_1^1 - a_2^2 + a_3^3 + 1}
\end{aligned} \tag{1.133}
$$

Notice that Eqs. (1.133) use only the diagonal elements of the transformation matrix and that the signs of the quaternion elements are not defined. A recommended method for the best numerical accuracy is to find the largest (in magnitude) element from Eqs. (1.133), use the positive sign, and then solve for the other elements. For instance, if q_1 were found to be the largest element,

$$
\begin{aligned}
q_2 &= (1/4)\left(a_3^2 - a_2^3\right)/q_1 \\
q_3 &= (1/4)\left(a_1^3 - a_3^1\right)/q_1 \\
q_4 &= (1/4)\left(a_2^1 - a_1^2\right)/q_1
\end{aligned} \tag{1.134}
$$

Similar equations can be written if q_2, q_3, or q_4 were found to have the largest numerical value from Eqs. (1.133). This method attaches the proper sign to the quaternion and distributes the errors into the quaternion that might arise due to the slight non-orthogonalities in the transformation matrix. Many aerospace vehicle motion simulations, as well as onboard guidance, navigation, and control (GNC) software systems, use this method to determine the attitude quaternion from the navigated vehicle body-axis-to-inertial transformation matrix.

1.4.6 Normalization and the Positive Quaternion

The definition of the Hamilton quaternion in Eq. (1.109), its transformation matrix of Eq. (1.111), and the fact that $q_1^2 + q_2^2 + q_3^2 + q_4^2 = 1.0$, all form the properties that assure that the transformation matrix will be orthogonal. However, after a quaternion has been computed from successive numerical processes or from a numerical integration process that will be discussed later in the text, normalization is recommended. This is done with the following computations:

$$|q| = \sqrt{q_1^2 + q_2^2 + q_3^2 + q_4^2} \tag{1.135}$$

and then

$$q_i' = q_i/|q| \tag{1.136}$$

where q_i' is the normalized quaternion q_i. This procedure simply evenly distributes the errors that may have accumulated in the computational process to each element of the quaternion and also guarantees an orthogonal transformation matrix. This procedure is necessary for computer algorithms, since computer accuracies are limited to some number of significant decimal digits.

It is also interesting to realize that when working with quaternions in computer computations, both the positive or the negative quaternion may result. It is shown in Eq. (1.113) that both quaternions are equal and will generate the same transformation matrix. The positive quaternion, i.e., where q_1 is positive, represents a right-hand rotation about the axis of rotation and is easier to visualize. However, during successive computations of the quaternion, such as numerical integrations, the sign of the first quaternion component q_1 should be allowed to change sign during the process to provide a continuous solution as q_1 passes through the 180 deg (π) angle.

1.4.7 Successive Coordinate Transformations Using the Quaternion and Quaternion Transformations

Consider the following coordinate transformations as described in Eqs. (1.50) and (1.51) and the resulting coordinate transformation of Eq. (1.52). Repeating these equations here for continuity of thought we have

$$x_i = a_i^\alpha \bar{x}_\alpha \tag{1.137}$$

$$\bar{x}_\alpha = b_\alpha^\gamma \bar{\bar{x}}_\gamma \tag{1.138}$$

and

$$x_i = a_i^\alpha b_\alpha^\gamma \bar{\bar{x}}_\gamma \tag{1.139}$$

The $\bar{\bar{\boldsymbol{x}}}$ vector components are transformed into the $\boldsymbol{x}$ vector by

$$c_i^\gamma = a_i^\alpha b_\alpha^\gamma \tag{1.140}$$

Again, Eq. (1.140) demonstrates the use of Cartesian matrix characteristics to develop each element of the c matrix as a function of the elements of both the a and b matrices. Likewise, the following equations for $(a) = (c)(b)^T$ and $(b) = (a)^T(c)$ can be written using the summation convention as follows:

$$a_i^\alpha = c_{i\gamma} b^{\alpha\gamma} \tag{1.141}$$

and

$$b_\alpha^\gamma = a_{\beta\alpha} c^{\beta\gamma} \tag{1.142}$$

It is very useful for future analyses to develop Eqs. (1.140), (1.141), and (1.142) as functions of the Hamilton quaternions that will define the equivalent quaternion multiply operations as these equations define the matrix multiply operations. Assume the following definitions:

1) q_ρ is the quaternion for matrix a,
2) p_β is the quaternion for matrix b, and
3) r_σ is the quaternion for matrix c.

Writing Eq. (1.141) as functions of the quaternions and using the same matrix structures from Eq. (1.115),

$$c_i^\gamma = R_i^{\gamma\sigma} r_\sigma = Q_i^{\alpha\rho} P_\alpha^{\gamma\beta} q_\rho p_\beta \tag{1.143}$$

Likewise, Eqs. (1.141) and (1.142) become

$$a_i^\alpha = Q_i^{\alpha\rho} q_\rho = R_{\gamma\sigma i} P^{\gamma\beta\alpha} r^\sigma p_\beta \tag{1.144}$$

and

$$b_\alpha^\gamma = P_\alpha^{\gamma\beta} p_\beta = Q_{\alpha\rho\delta} R^{\gamma\sigma\delta} r_\sigma q^\rho \tag{1.145}$$

In the interest of a standardized notation in the text, as alluded to earlier in Eq. (1.115), the author suggests the use of the superscripts and subscripts representing the ith rows, jth columns, and kth planes be defined as follows:

$$Q_k^{ij} = Q^{ijk} = Q_{ijk}$$

or

$$Q_{(\text{planes})}^{(\text{rows})(\text{columns})} = Q^{(\text{rows})(\text{columns})(\text{planes})} = Q_{(\text{rows})(\text{columns})(\text{planes})} \tag{1.146}$$

If the quaternions q_ρ and p_β are known, the c matrix is also known from Eq. (1.143), but the quaternion r_σ would have to be extracted from the c matrix using the quaternion extraction process as discussed in Sec. 1.4.5. Using quaternion algebra, the literature discussed in Refs. 11 and 15 shows that a direct transformation exists of the form

$$r_\sigma = \mathcal{Q}_\sigma^\beta p_\beta \tag{1.147}$$

where the $\mathcal{Q}_\sigma^\beta$ again uses calligraphic capital letter notation to represent a 4×4 matrix whose elements are formed with the q_ρ quaternion. Furthermore, it is shown that the $\mathcal{Q}$ matrix is orthogonal, and so we can write

$$p_\beta = \mathcal{Q}_{\sigma\beta} r^\sigma \tag{1.148}$$

The existence of the $\mathcal{Q}$ matrix may be demonstrated with the following analysis starting with the subsitution of Eq. (1.147) into Eq. (1.145):

$$\left(P_\alpha^{\gamma\beta}\right)p_\beta = \left(\mathcal{Q}_{\alpha\rho\delta}R^{\gamma\sigma\delta}q^\rho \mathcal{Q}_\sigma^\beta\right)p^\beta \tag{1.149}$$

and equating factors of p_β, we have

$$P_\alpha^{\gamma\beta} = \mathcal{Q}_{\alpha\rho\delta}R^{\gamma\sigma\delta}q^\rho \mathcal{Q}_\sigma^\beta \tag{1.150}$$

Since the $P_\alpha^{\gamma\beta}$ structure is the same as shown in Eq. (1.115), there are 36 equalities that can be written from Eq. (1.150). By selecting, for instance, three of these equalities, namely $P_1^{11} = P_1^{33}$, $P_2^{14} = P_2^{21}$, and $P_3^{22} = P_3^{31}$, the structure of the $\mathcal{Q}$ matrix is revealed to be

$$(\mathcal{Q}) = \begin{pmatrix} \mathcal{Q}_1^1 & -\mathcal{Q}_2^1 & -\mathcal{Q}_3^1 & -\mathcal{Q}_4^1 \\ \mathcal{Q}_2^1 & \mathcal{Q}_1^1 & -\mathcal{Q}_4^1 & \mathcal{Q}_3^1 \\ \mathcal{Q}_3^1 & \mathcal{Q}_4^1 & \mathcal{Q}_1^1 & -\mathcal{Q}_2^1 \\ \mathcal{Q}_4^1 & -\mathcal{Q}_3^1 & \mathcal{Q}_2^1 & \mathcal{Q}_1^1 \end{pmatrix} \tag{1.151}$$

Continuing, we substitute Eq. (1.148) into Eq. (1.143) and equating factors of p_β, we have

$$R_i^{\gamma\sigma} \mathcal{Q}_\sigma^\beta = \mathcal{Q}_i^{\alpha\rho} P_\alpha^{\gamma\beta} q_\rho \tag{1.152}$$

By expanding this equation and using the constraints as defined by the structure of the $\mathcal{Q}$ matrix in Eq. (1.151),we have $\mathcal{Q}_1^1 = q_1$, $\mathcal{Q}_2^1 = q_2$, $\mathcal{Q}_3^1 = q_3$, and $\mathcal{Q}_4^1 = q_4$, or simply elements of the q_ρ quaternion! Using this relationship, Eq. (1.147) becomes

$$\begin{pmatrix} r_1 \\ r_2 \\ r_3 \\ r_4 \end{pmatrix} = \begin{pmatrix} q_1 & -q_2 & -q_3 & -q_4 \\ q_2 & q_1 & -q_4 & q_3 \\ q_3 & q_4 & q_1 & -q_2 \\ q_4 & -q_3 & q_2 & q_1 \end{pmatrix} \begin{pmatrix} p_1 \\ p_2 \\ p_3 \\ p_4 \end{pmatrix} \tag{1.153}$$

This relationship is much more easily derived using quaternion algebra as described in Sec. 1.4.4 of the text, with the following analysis. The corresponding quaternion equations, analogous to the tensor Eqs. (1.137), (1.138), and (1.139), can be written

$$\tilde{x} = \tilde{q}\tilde{\tilde{x}}\tilde{q}^* \tag{1.154}$$

$$\tilde{\tilde{x}} = \tilde{p}\tilde{\tilde{\tilde{x}}}\tilde{p}^* \tag{1.155}$$

and

$$\tilde{x} = \tilde{r}\tilde{\tilde{\tilde{x}}}\tilde{r}^* \tag{1.156}$$

respectively. Substituting Eq. (1.155) into Eq. (1.154), we have

$$\tilde{x} = (\tilde{q}\tilde{p})\tilde{\tilde{\tilde{x}}}(\tilde{p}^*\tilde{q}^*) \tag{1.157}$$

Comparing Eqs. (1.157) and (1.156), we have the desired quaternion algebraic relationship

$$\tilde{r} = \tilde{q}\tilde{p}$$

or

$$\tilde{r} = (q_1 + q_2 i + q_3 j + q_4 k)(p_1 + p_2 i + p_3 j + p_4 k) \tag{1.158}$$

and carrying out the indicated operations, we have again Eq. (1.153).

The transformation equation for q_ρ as functions of the p and r quaternions can also be demonstrated in a manner similar to the preceding analysis by starting with

$$q_\rho = \mathcal{R}_\rho^\beta p_\beta \tag{1.159}$$

The $\mathcal{R}$ matrix has a slightly different element sign structure than the $\mathcal{Q}$ matrix and is as follows:

$$(\mathcal{R}) = \begin{pmatrix} r_1 & r_2 & r_3 & r_4 \\ r_2 & -r_1 & r_4 & -r_3 \\ r_3 & -r_4 & -r_1 & r_2 \\ r_4 & r_3 & -r_2 & -r_1 \end{pmatrix} \tag{1.160}$$

Hence a set of transformation equations are defined for the quaternions. The quaternion transformation equations can be used in the same manner corresponding to the matrix multiply operations being performed. These corresponding equations are summarized as follows:

$$(c) = (a)(b): \qquad r_\sigma = \mathcal{Q}_\sigma^\beta p_\beta = \mathcal{P}_\sigma^\rho q_\rho \tag{1.161}$$

$$(a) = (c)(b)^T: \qquad q_\rho = \mathcal{R}_\rho^\beta p_\beta = \mathcal{P}_{\sigma\rho} r^\sigma \tag{1.162}$$

and

$$(b) = (a)^T(c): \qquad p^\beta = \mathcal{Q}_\sigma^\beta r^\sigma = \mathcal{R}_\rho^\beta q^\rho \tag{1.163}$$

1.4.8 Aerospace Applications: Space Vehicle Steering Using the Steering Quaternion

Aircraft or space vehicle onboard control functions are divided logically into four distinct systems groups, each requiring specific analysis disciplines to effectively control and fly the vehicle in the designed flight envelope. These functional groups may be visualized in Fig. 1.13. The onboard control loop shown in the figure must be cycled rapidily enough to accurately control the flight of the vehicle. For instance, a cycle time of 25 times per second or 25 Hz (where one hertz is one cycle per second, named after Heinrich R. Hertz in 1928) may be required to control the rapid changes in vehicle attitude. Each of these control functions may be thought of as follows:

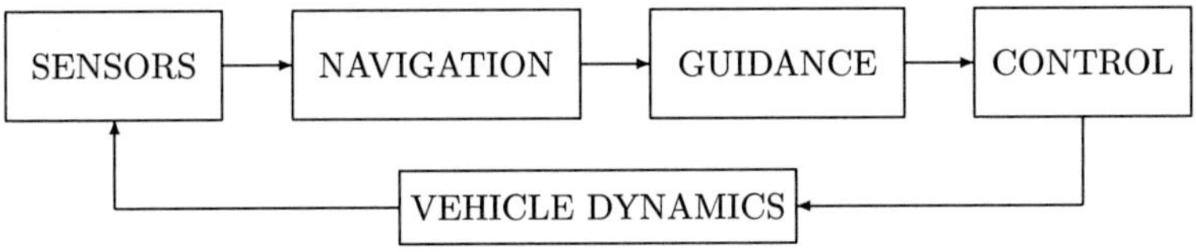

Fig. 1.13 The GNC control functions.

1) Sensors: the physical measurement of the pertinent vehicle motion parameters.

2) Navigation: using the measurements from the vehicle's onboard sensors to compute the actual motion states of the vehicle, for example, position, velocity, acceleration, attitude, and vehicle body axis rotation rates.

3) Guidance: computes the desired vehicle motion states to fly the vehicle along the designed mission profile.

4) Control: using the difference between the desired vehicle states and the vehicle actual states, activate the appropriate vehicle control effectors (aerodynamic control surface deflection angles, throttle settings, rocket engine gimbal control angles, etc.) to fly the vehicle to the desired flight profile.

In Sec. 1.2.6 the star tracker is the sensor providing the measurements to the navigation function that determines the space vehicle attitude relative to the ECI reference frame. Likewise, in Sec. 1.3.6 the inertial measurement unit becomes the sensor unit providing angles to the navigation system to compute the vehicle body to ECI transformation matrix to estimate actual vehicle attitude. The actual attitude navigation need not necessarily be referenced to the ECI frame, however, in this application it is instructive to use coordinates that have been previously discussed.

For this example, therefore, let the actual body to ECI transformation matrix and its quaternion from the space vehicle's sensor and navigation systems be represented by Eq. (1.137) or simply the matrix a and its quaternion q:

$$\boldsymbol{x} = (a)\bar{\boldsymbol{x}} \quad \text{(actual body to ECI, with quaternion } q\text{)} \tag{1.164}$$

Further, the space vehicle's guidance system as shown in Fig. 1.13 computes the desired attitude in the inertial frame (the ECI frame) that can be represented by the matrix c from Eq. (1.139) and (1.140) with its associated quaternion r:

$$\boldsymbol{x} = (c)\bar{\bar{\boldsymbol{x}}} \quad \text{(desired body to ECI, with quaternion } r\text{)} \tag{1.165}$$

By equating Eqs. (1.164) and (1.165), the b matrix as in Eq. (1.138) is formed that transforms vectors in the desired body coordinates into vectors in actual body coordinates:

$$\bar{\boldsymbol{x}} = (a)^T(c)\bar{\bar{\boldsymbol{x}}} \tag{1.166}$$

The quaternion p, from $(a)^T(c)$ as well as this matrix, as shown in Fig. 1.12, rotates the actual body coordinate frame (in this example the $\bar{x}$ frame) into the desired coordinate frame (here the $\bar{\bar{x}}$ frame). The b matrix, namely $(b) = (a)^T(c)$, as in

Eq. (1.142) or the quaternion operation of Eq. (1.163), yields the quaternion p, which is referred to as the *steering quaternion*.[16] The steering quaternion from Eqs. (1.109) gives the amount of rotation ω that must occur and the eigenaxis (the direction angles of α, β, and γ) around which the body axes must be rotated. This can be visualized using Fig. 1.14.

The steering quaternion can be used as follows to determine each body axis rotation rate command necessary for the flight control system function, shown in Fig. 1.13, to activate the vehicle's control effectors and cause the vehicle to rotate to the desired attitude.

For instance, if ω approaches zero, i.e., $p_1 = 1$, the *actual-body-axis*-to-*desired-body-axis* transformation matrix becomes the unit matrix, as seen from Eq. (1.109), and no control action may be required.

If ω is not equal to one (within some tolerance), the vehicle's control system will command the vehicle to rotate about the eigenaxis defined by the vector part of the steering quaternion (p_2, p_3, p_4) by the amount ω. For example, on each control cycle the flight control system will determine an allowable closure rate, $\bar{\omega}_{\max}$, to control and reduce the angle ω to zero. The body axis rotation rate commands are computed as follows:

$$
\begin{aligned}
(\bar{\omega}_1)_c &= \bar{\omega}_{\max}\left(\frac{p_2}{\sin(1/2)\omega}\right)\\
(\bar{\omega}_2)_c &= \bar{\omega}_{\max}\left(\frac{p_3}{\sin(1/2)\omega}\right)\\
(\bar{\omega}_3)_c &= \bar{\omega}_{\max}\left(\frac{p_4}{\sin(1/2)\omega}\right)
\end{aligned}
\tag{1.167}
$$

Please note here that the $\bar{\omega}$ variables refer to body axis rotation rates, which will be discussed in detail in Sec. 1.5.2, and should not be confused with the angular measurement ω as shown in Fig. 1.12 and used in Eqs. (1.109).

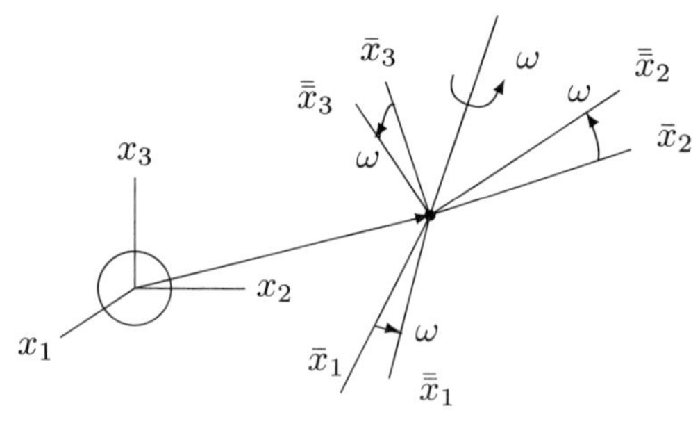

Fig. 1.14 The steering quaternion.

1.5 Differentiation of the Transformation Matrix

1.5.1 Time Derivative of the Position Vector

Rewriting Eq. (1.14) in the form

$$x_i = a_i^\alpha \bar{x}_\alpha + X_i \tag{1.168}$$

and differentiating with respect to time, we have

$$\frac{\mathrm{d}x_i}{\mathrm{d}t} = \left(\frac{\mathrm{d}a_i^\alpha}{\mathrm{d}t}\right)\bar{x}_\alpha + a_i^\alpha\left(\frac{\mathrm{d}\bar{x}_\alpha}{\mathrm{d}t}\right) + \frac{\mathrm{d}X_i}{\mathrm{d}t}$$

or simply,

$$v_i = \left(\frac{\mathrm{d}a_i^\alpha}{\mathrm{d}t}\right)\bar{x}_\alpha + a_i^\alpha \bar{v}_\alpha + V_i \tag{1.169}$$

where the time derivatives of position components are the velocity vector components, i.e., v_i, $\bar{v}_\alpha$, and V_i. For practical applications, further analysis is needed to define the time derivative of the transformation matrix. Rewriting Eq. (1.37) and realizing that $(a)^{-1} = (a)^T$ for orthogonal matrices, we have

$$(a)^T(a) = (a)(a)^T = I \tag{1.170}$$

By differentiating this equation, we can write

$$(\dot{a})^T(a) + (a)^T(\dot{a}) = (\dot{a})(a)^T + (a)(\dot{a})^T = 0 \tag{1.171}$$

where the dots indicate differentiation with respect to time. Using just the left-hand side of the equalities of Eq. (1.171) we have

$$(a)^T(\dot{a}) = -(\dot{a})^T(a)$$

or

$$(a)^T(\dot{a}) = -((a)^T(\dot{a}))^T \tag{1.172}$$

A matrix that is equal to the negative of its transpose defines a special kind of matrix called a skew-symmetric matrix (Ref. 6, Sec. 5–6). We can write Eq. (1.172) simply as $(a)^T(\dot{a}) = (\bar{W})$, where $\bar{W}$ is some skew-symmetric matrix and solving for $\dot{a}$, we have

$$(\dot{a}) = (a)(\bar{W}) \tag{1.173}$$

From this analysis, we can conclude that the derivative of a transformation matrix is equal to itself times a skew-symmetric matrix. However, the elements of the $\bar{W}$ matrix need to be defined.

1.5.2 *Axis Rotation Rates and the Derivative of the Transformation Matrix*

Consider at time t, that there is some transformation matrix such that

$$\boldsymbol{x} = (a)\bar{\boldsymbol{x}} \tag{1.174}$$

which is illustrated in Fig. 1.15. A very short time later, say, $t + \delta t$ (where δt is a small variation in time), the $\bar{x}$ coordinate frame axes are displaced and are represented by the $\bar{\bar{x}}$ coordinate frame (Ref. 13, Sec. 4–7). Using the arc length relationship, $\delta s = r\delta\theta$, where r is the unit vector length, the new coordinate axes with respect to $\bar{x}$ are given by

$$\begin{aligned}
\bar{\bar{x}}_1 &= 1\bar{\boldsymbol{i}} + \delta\theta_3\bar{\boldsymbol{j}} - \delta\theta_2\bar{\boldsymbol{k}} \\
\bar{\bar{x}}_2 &= -\delta\theta_3\bar{\boldsymbol{i}} + 1\bar{\boldsymbol{j}} + \delta\theta_1\bar{\boldsymbol{k}} \\
\bar{\bar{x}}_3 &= \delta\theta_2\bar{\boldsymbol{i}} - \delta\theta_1\bar{\boldsymbol{j}} + 1\bar{\boldsymbol{k}}
\end{aligned} \tag{1.175}$$

where the $\bar{\boldsymbol{i}}, \bar{\boldsymbol{j}}$, and $\bar{\boldsymbol{k}}$ are the unit vectors along the $\bar{x}$ axes. The $\delta\theta_1$, $\delta\theta_2$, and $\delta\theta_3$ are the small rotations about the coordinate axes $\bar{x}_1$, $\bar{x}_2$, and $\bar{x}_3$, respectively. These rotations are made using the right-hand rule, i.e., with the thumb of the right hand extended, the curled fingers show the direction of the rotation. The $\bar{x}$ axes move to the $\bar{\bar{x}}$ axes positions given by Eqs. (1.175) and may be visualized with the help of Fig. 1.16.

Using the properties of the transformation matrix described in Eqs. (1.21), the transformation matrix that transforms vectors in the $\bar{\bar{x}}$ into the $\bar{x}$ frame becomes

$$(b) = (\bar{\bar{x}}_1, \bar{\bar{x}}_2, \bar{\bar{x}}_3) = \begin{pmatrix} 1 & -\delta\theta_3 & \delta\theta_2 \\ \delta\theta_3 & 1 & -\delta\theta_1 \\ -\delta\theta_2 & \delta\theta_1 & 1 \end{pmatrix} \tag{1.176}$$

where

$$\bar{\boldsymbol{x}} = (b)\bar{\bar{\boldsymbol{x}}} \tag{1.177}$$

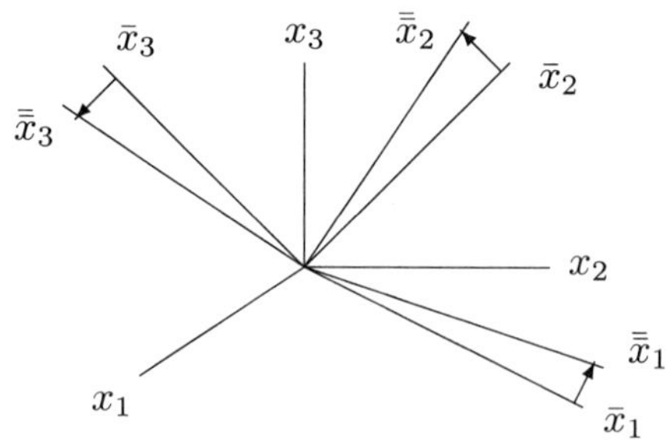

Fig. 1.15 Small angle displacements.

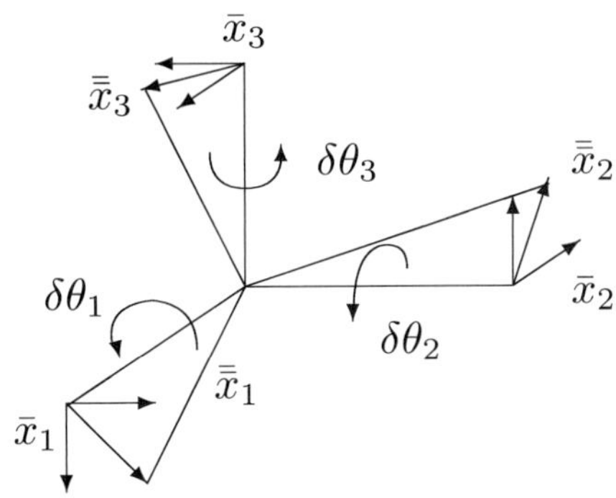

Fig. 1.16 Small angle per axis movement.

The transformation to vectors in the x coordinate frame at time $t + \delta t$ from Eqs. (1.174) and (1.177) becomes

$$\boldsymbol{x} = (a)(b)\bar{\bar{\boldsymbol{x}}} \tag{1.178}$$

The definition of the derivative of the transformation matrix is

$$\frac{\mathrm{d}a}{\mathrm{d}t} = \lim_{\delta t \to 0} \left(\frac{a(t + \delta t) - a(t)}{\delta t} \right) \tag{1.179}$$

The notation here is simpified where $a(t + \delta t)$ represents the transformation matrix at time $t + \delta t$ and $a(t)$ is transformation matrix at time t. Substituting from Eqs. (1.174) and (1.178), we have

$$\frac{\mathrm{d}a}{\mathrm{d}t} = \lim_{\delta t \to 0} \left(\frac{(a)(b) - (a)}{\delta t} \right) = \lim_{\delta t \to 0} \left(\frac{(a)((b) - (I))}{\delta t} \right)$$

and finally,

$$\frac{\mathrm{d}a}{\mathrm{d}t} = (a) \lim_{\delta t \to 0} \left(\frac{(b) - (I)}{\delta t} \right) \tag{1.180}$$

From Eq. (1.176) the expression $((b) - (I))/\delta t$ becomes

$$\frac{1}{\delta t} \begin{pmatrix} 0 & -\delta\theta_3 & \delta\theta_2 \\ \delta\theta_3 & 0 & -\delta\theta_1 \\ -\delta\theta_2 & \delta\theta_1 & 0 \end{pmatrix}$$

and the last equation of Eq. (1.180) is

$$\frac{\mathrm{d}a}{\mathrm{d}t} = (a) \lim_{\delta t \to 0} \begin{pmatrix} 0 & -\left(\dfrac{\delta\theta_3}{\delta t}\right) & \left(\dfrac{\delta\theta_2}{\delta t}\right) \\ \left(\dfrac{\delta\theta_3}{\delta t}\right) & 0 & -\left(\dfrac{\delta\theta_1}{\delta t}\right) \\ -\left(\dfrac{\delta\theta_2}{\delta t}\right) & \left(\dfrac{\delta\theta_1}{\delta t}\right) & 0 \end{pmatrix} \tag{1.181}$$

where

$$\lim_{\delta t \to 0} \left(\frac{\delta \theta_n}{\delta t} \right) = \bar{\omega}_n$$

are simply the coordinate axis rotation rates, and finally in matrix form we have

$$\frac{\mathrm{d}a}{\mathrm{d}t} = (a) \begin{pmatrix} 0 & -\bar{\omega}_3 & \bar{\omega}_2 \\ \bar{\omega}_3 & 0 & -\bar{\omega}_1 \\ -\bar{\omega}_2 & \bar{\omega}_1 & 0 \end{pmatrix} = (a)(\bar{W}) \tag{1.182}$$

This expression verifies Eq. (1.173) and also defines the elements of the skew-symmetric-coordinate-axis-rotation-rate matrix $\bar{W}$. Using the summation notation for the component form, Eq. (1.182) can be written

$$\dot{a}_i^\alpha = a_i^\gamma \bar{W}_\gamma^\alpha \tag{1.183}$$

Later in the text when coordinate and body dynamics are discussed, the coordinate axes define the space vehicle body axes and $\bar{W}$ becomes the body-axis-rotation-rate matrix.

1.5.3 Euler Angles Rates and the Derivative of the Transformation Matrix

When the transformation matrix is written as a function of the Euler angles, then its derivative becomes

$$\frac{\mathrm{d}a_i^\alpha}{\mathrm{d}t} = \frac{\mathrm{d}a_i^\alpha(\theta_1, \theta_2, \theta_3)}{\mathrm{d}t} = \frac{\partial a_i^\alpha}{\partial \theta_\beta} \frac{\mathrm{d}\theta_\beta}{\mathrm{d}t} \tag{1.184}$$

where θ_1, θ_2, and θ_3 are the three Euler angles used in a desired rotational sequence to describe a particular problem. The partial derivatives of the elements of the transformation matrix may be written

$$B_i^{\alpha\beta} = \frac{\partial a_i^\alpha}{\partial \theta_\beta}$$

so that Eq. (1.184) becomes

$$\frac{\mathrm{d}a_i^\alpha}{\mathrm{d}t} = B_i^{\alpha\beta} \dot{\theta}_\beta \tag{1.185}$$

This equation defines the rate of change of the transformation matrix as a function of the rates of change of the Euler angles and may be equated to Eq. (1.183):

$$\dot{a}_i^\alpha = B_i^{\alpha\beta} \dot{\theta}_\beta = a_i^\gamma \bar{W}_\gamma^\alpha \tag{1.186}$$

This equation forms an angular velocity equality from which the relationships between the rates of change of the Euler angles, $\dot{\theta}_\beta$, and the coordinate axes rotation rates, $\bar{\omega}_\rho$, can be derived. The $\dot{\theta}_\beta$ are measured from the moving Euler angles being used in the analysis and are therefore different for each of the 12 Euler angle sequences. The coordinate angular velocities $\bar{\omega}_i$, on the other hand, are measured along each coordinate axis and are represented by a vector in the rotating frame with components along each of the respective rotating axes. This vector can be transformed into the stationary frame (the unbarred frame) the same as a position vector by

$$\omega_i = a_i^\alpha \bar{\omega}_\alpha \tag{1.187}$$

We can likewise extend the angular velocity equality of Eq. (1.186) as functions of the angular velocity components in the unbarred frame (in the inertial frame) using Eq. (1.187). This equality is not generally as useful in rotating dynamics since most of the measureable quantities involve the rotating axes. Nevertheless, the angular velocity equality can be extended to

$$\dot{a}_i^\alpha = B_i^{\alpha\beta}\dot{\theta}_\beta = a_i^\gamma \bar{W}_\gamma^\alpha = a_\rho^\alpha W_i^\rho \tag{1.188}$$

where

$$W_i^\rho = \begin{pmatrix} 0 & -\omega_3 & \omega_2 \\ \omega_3 & 0 & -\omega_1 \\ -\omega_2 & \omega_1 & 0 \end{pmatrix} \tag{1.189}$$

1.5.4 Axis Rotation Rates and Rate of Change of the Euler Angles

In many problems of dynamics involving rotating Cartesian coordinate systems, both the coordinate axis rotation rates and the Euler angle rates are employed to describe the motion. For this reason it is necessary to develop the relationships that will give the $\dot{\theta}_1$, $\dot{\theta}_2$, and $\dot{\theta}_3$ and sometimes their derivatives as functions of the coordinate axis (body axis) rotation rates and their derivatives, respectively. By examining Fig. 1.17, which depicts the standard aircraft yaw-pitch-roll Euler sequence, we can analyze the relationships between the coordinate axis (body axis) rotation rates and the rates of change of the Euler angles.

The total instantaneous angular velocity in the unbarred frame can be deduced from the geometry of Fig. 1.17 and can be written

$$\mathbf{\Omega} = \dot{\Psi}\boldsymbol{k} + \dot{\Theta}\boldsymbol{n} + \dot{\Phi}\bar{\boldsymbol{i}} \tag{1.190}$$

The angles Ψ, Θ, and Φ, are standard aeronautical notation used for the aircraft or spacecraft yaw, pitch, and roll Euler angles, respectively. However, for consistency in the notation for representation in equations, we rewrite Eq. (1.190) as

$$\mathbf{\Omega} = \dot{\theta}_1\boldsymbol{k} + \dot{\theta}_2\boldsymbol{n} + \dot{\theta}_3\bar{\boldsymbol{i}} \tag{1.191}$$

where $\dot{\theta}_1 = \dot{\Psi}$, $\dot{\theta}_2 = \dot{\Theta}$, and $\dot{\theta}_3 = \dot{\Phi}$, again in comparison with the standard aeronautical notation. The vectors $\boldsymbol{i}, \boldsymbol{j}$, and $\boldsymbol{k}$ are along the unbarred (inertial) axes.

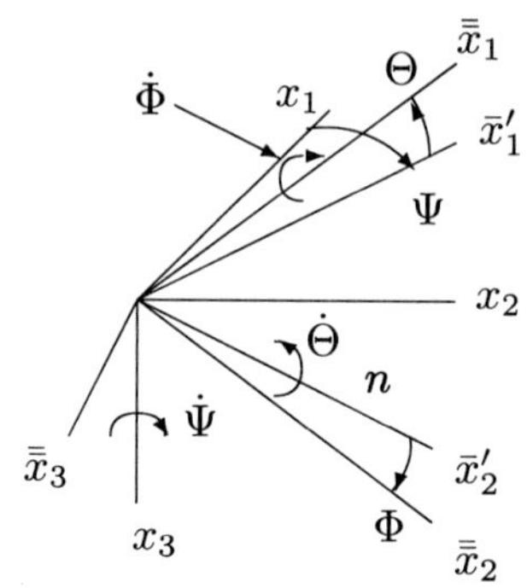

Fig. 1.17 Euler angle rotation rates.

The vectors $\bar{\boldsymbol{i}}, \bar{\boldsymbol{j}}$, and $\bar{\boldsymbol{k}}$ are directed along the moving axes and are defined instantaniously in the inertial coordinate frame by the elements of the transformation matrix. The instantaneous Euler pitch rate occurs about the vector $\boldsymbol{n}$, which is the same as the intermediate $\bar{x}_2'$ axis; $\boldsymbol{n}$ is simply $\boldsymbol{n} = -\sin\theta_1 \boldsymbol{i} + \cos\theta_1 \boldsymbol{j}$ as defined in this 3-2-1 Euler sequence of rotations. The total angular velocity vector about the moving axes, $\bar{x}_1$, $\bar{x}_2$, and $\bar{x}_3$, is

$$\boldsymbol{\Omega} = \bar{\omega}_1 \bar{\boldsymbol{i}} + \bar{\omega}_2 \bar{\boldsymbol{j}} + \bar{\omega}_3 \bar{\boldsymbol{k}} \tag{1.192}$$

As described in Sec. 1.1, Eq. (1.8), and since Eqs. (1.191) and (1.192) describe the same coordinate rotation, $\boldsymbol{\Omega}$ becomes invariant as viewed from either coordinate frame. Therefore, we can form the vector dot products of Eqs. (1.191) and (1.192) with $\bar{\boldsymbol{i}}, \bar{\boldsymbol{j}}$, and $\bar{\boldsymbol{k}}$, one at a time, and equate the results of the operations on each equation. Then using Eqs. (1.11) and the coordinate transformation matrix given in Appendix A.10, we have

$$\begin{aligned}
\bar{\omega}_1 &= -\sin\theta_2 \dot{\theta}_1 + \dot{\theta}_3 \\
\bar{\omega}_2 &= \cos\theta_3 \dot{\theta}_2 + \sin\theta_3 \cos\theta_2 \dot{\theta}_1 \\
\bar{\omega}_3 &= -\sin\theta_3 \dot{\theta}_2 + \cos\theta_3 \cos\theta_2 \dot{\theta}_1
\end{aligned} \tag{1.193}$$

Using the standard aeronautical notation for the yaw-pitch-roll Euler angles and their rates of change, Eq. (1.193) is written

$$\begin{aligned}
p &= -\sin\Theta \dot{\Psi} + \dot{\Phi} \\
q &= \cos\Phi \dot{\Theta} + \sin\Phi \cos\Theta \dot{\Psi} \\
r &= -\sin\Phi \dot{\Theta} + \cos\Phi \cos\Theta \dot{\Psi}
\end{aligned} \tag{1.194}$$

where p, q, and r are the body axis rotation rates. Solving these equations simultaneously for $\dot{\Psi}$, $\dot{\Theta}$, and $\dot{\Phi}$, we have the relationship

$$\begin{aligned}
\dot{\Psi} &= (\sin\Phi q + \cos\Phi r)/\cos\Theta \\
\dot{\Theta} &= \cos\Phi q - \sin\Phi r \\
\dot{\Phi} &= p + (\sin\Phi \sin\Theta q + \cos\Phi \sin\Theta r)/\cos\Theta
\end{aligned} \tag{1.195}$$

A singular point appears when the Euler angles are used to describe rotational motion, as pointed out in Sec. 1.3.4. This singular point is noted here in Eqs. (1.195) when Θ approaches 90 deg the cosine approaches zero. Therefore, the use of Eqs. (1.195) in computer applications must be limited in numerical calculations in the region where Θ is near 90 deg.

1.5.5 Recursive Relationships for Coordinate Axis Rotation Rates

There are a number of Euler angle sequences (12 possible sets for three Euler rotations as described in Sec. 1.3.3), any one of which can be used to describe a problem or may be required by hardware definition. As shown in Sec. 1.5.4, each equation relating the body (coordinate) axis rotation rates to the Euler angle rates of change is unique for that sequence. These relationships, as those previously described for the 3-2-1 Euler sequence, are not generally written out for aerospace engineering applications. The following analysis, as suggested by Corben and Stehle (Ref. 9, p. 143) and as described in Ref. 18, provides a method of computing all of these relationships and extends the equations to include n-Euler rotations. This method lends itself immediately to computer application and can also be used to write out the defining Euler rate relationships when a particular Euler sequence has been selected.

If we differentiate Eq. (1.94), we can write

$$\begin{aligned}
\dot{a} &= \dot{R}_1 R_2 R_3 \cdots R_n + R_1 \dot{R}_2 R_3 \cdots R_n \\
&\quad + R_1 R_2 \dot{R}_3 \cdots R_n + \cdots + R_1 R_2 R_3 \cdots \dot{R}_n
\end{aligned} \tag{1.196}$$

Since each of the R_n are single-axis transformation matrices, their derivatives are given by Eq. (1.183) and can be written

$$\dot{R}_n = R_n W_n \tag{1.197}$$

where the W_n, in this analysis, are the skew-symmetric matrices containing only the Euler rate about the nth axis for that rotation sequence. For example, if the nth Euler rotation were about the y axis (the pitch axis) and we refer to Eq. (1.182), we have

$$W_2 = \begin{pmatrix} 0 & 0 & \dot{\Theta} \\ 0 & 0 & 0 \\ -\dot{\Theta} & 0 & 0 \end{pmatrix} \tag{1.198}$$

where $\dot{\Theta}$ is the pitch rate. Substituting Eq. (1.197) into Eq. (1.196), the following recursive form results:

$$a^T \dot{a} = \bar{U}_n = \bar{W} \tag{1.199}$$

where

$$\bar{U}_n = R_n^T \bar{U}_{n-1} R_n + W_n \tag{1.200}$$

and defining $\bar{U}_0 = 0$, for the starting solution. The notation $\bar{U}_n$ means a matrix formed as a result of n Euler rotations. Equation (1.200) provides the analytic solution for the elements of the body-axis-rotation-rate matrix $\bar{W}$, as functions of the single-axis Euler rotation matrices and the Euler angle rates of change. For example, when $n = 3$, any three Euler rotational sequences, we have $\bar{W} = \bar{U}_3$ as follows:

$$\bar{U}_3 = R_3^T \left(R_2^T W_1 R_2 + W_2\right) R_3 + W_3 \tag{1.201}$$

which was derived from Eq. (1.200) in the following manner:

$$\begin{aligned}
\bar{U}_1 &= R_1^T \bar{U}_0 R_1 + W_1 = W_1 \\
\bar{U}_2 &= R_2^T \bar{U}_1 R_2 + W_2 = R_2^T W_1 R_2 + W_2 \\
\bar{U}_3 &= R_3^T \bar{U}_2 R_3 + W_3 = R_3^T \left(R_2^T W_1 R_2 + W_2\right) R_3 + W_3
\end{aligned} \tag{1.202}$$

Notice in the preceding analysis each coordinate (body) axis rotation rate of change is solved for twice, as seen in the $\bar{W}$ matrix Eq. (1.182). The following analysis uses the vector characteristics of the Euler angle rates of change and is somewhat easier to use when writing the equations defining the components of the body axis rotation rate vector.

The Euler angles are measured from the inertial (stationary) coordinate frame to the instantaneous Euler axis position in the particular sequence that is being used. The rate of change of the Euler angles refers to rotations about one axis at a time, using the intermediate Euler axes as defined by the selected sequence. For instance, in the yaw-pitch-roll (3-2-1) Euler sequence, the rate of change of the pitch angle is measured about a line where the pitch is occurring (the $\bar{x}_2'$ axis as shown in Fig. 1.17) and creates components of the rotation vector in the inertial frame when only the first two rotations have occurred. The resulting inertial vector components from the second rotation are

$$\omega_2' = R_1 R_2 \begin{pmatrix} 0 \\ \dot{\Theta} \\ 0 \end{pmatrix} \tag{1.203}$$

where $R_1 = Z(\Psi)$, a rotation about the yaw axis (a z-axis rotation), and $R_2 = Y(\Theta)$, a rotation about the pitch axis (a y-axis rotation). The total inertial rotation vector is the sum of all rotation components resulting from the rates of change of each Euler angle, that is,

$$\omega = \omega_1' + \omega_2' + \omega_3' \tag{1.204}$$

for a three-axis Euler sequence. Writing out the sum resulting from n-Euler angle rates of change

$$\boldsymbol{\omega} = R_1\boldsymbol{S}_1 + R_1R_2\boldsymbol{S}_2 + R_1R_2R_3\boldsymbol{S}_3 + \cdots + R_1R_2R_3\cdots R_n\boldsymbol{S}_n \tag{1.205}$$

The $\boldsymbol{S}_n$ vectors are the instantaneous Euler rate vectors about the particular axis selected by the Euler angle sequence as shown in the example Eq. (1.203). Equation (1.205) is the inertial rotation rate vector (about the stationary axes), however, the body axis or moving axis rotation rates are of primary interest and find the most application in aerospace problems. The inertial rotation rate vector is transformed to the moving coordinate axes by simply $\bar{\boldsymbol{\omega}} = a^T\boldsymbol{\omega}$. Substituting $\boldsymbol{\omega}$ from Eq. (1.205) and forming a^T from Eq. (1.94), we can write

$$\bar{\boldsymbol{\omega}} = \left(R_n^T\cdots R_3^TR_2^TR_1^T\right)(R_1\boldsymbol{S}_1 + R_1R_2\boldsymbol{S}_2 + R_1R_2R_3\boldsymbol{S}_3 + \cdots + R_1R_2R_3\cdots R_n\boldsymbol{S}_n)$$

which reduces to

$$\bar{\boldsymbol{\omega}} = R_n^T\cdots R_3^TR_2^T\boldsymbol{S}_1 + R_n^T\cdots R_4^TR_3^T\boldsymbol{S}_2 + R_n^T\cdots R_5^TR_4^T\boldsymbol{S}_3 + \cdots + \boldsymbol{S}_n$$

This equation may be written in recursive form as follows:

$$\bar{\boldsymbol{\omega}} = \bar{\boldsymbol{T}}_n \tag{1.206}$$

where the $\bar{\boldsymbol{T}}_n$ vector has the form

$$\bar{\boldsymbol{T}}_n = R_n^T\bar{\boldsymbol{T}}_{n-1} + \boldsymbol{S}_n \tag{1.207}$$

again defining $\bar{\boldsymbol{T}}_0 = 0$ for the starting solution. The notation $\bar{\boldsymbol{T}}_n$ means a vector formed as a result of n Euler rotations, where n is the number of Euler rotations, R_n is the nth single-axis rotation matrix and $\boldsymbol{S}_n$ is a vector of the nth Euler angle time derivative. For example, when $n = 3$, any three Euler rotational sequences, Eq. (1.206) reduces to

$$\bar{\boldsymbol{\omega}} = R_3^T\left(R_2^T\boldsymbol{S}_1 + \boldsymbol{S}_2\right) + \boldsymbol{S}_3 \tag{1.208}$$

the body axis (the moving coodinate frame) rotation rate vector as a function of the Euler angle rates of change and the single-axis Euler rotation matrices.

1.5.6 Quaternion Rates and the Derivative of the Transformation Matrix

When the transformation matrix is a function of the Hamilton quaternion as in Eq. (1.111), its derivative can be written

$$\frac{\mathrm{d}a_i^\alpha}{\mathrm{d}t} = \frac{\partial a_i^\alpha}{\partial q_\beta}\dot{q}^\beta \tag{1.209}$$

By differentiating Eqs. (1.115), we can also write

$$\frac{\mathrm{d}a_i^\alpha}{\mathrm{d}t} = 2Q_i^{\alpha\rho}\dot{q}_\rho \tag{1.210}$$

Likewise, differentiating Eq. (1.116), we have

$$\frac{da_i^\alpha}{dt} = 2S_i^\beta \dot{T}_\beta^\alpha = 2\dot{S}_i^\beta T_\beta^\alpha \tag{1.211}$$

These relationships involve 3 × 4 matrices and are difficult to work with but may find applications in some analyses. Therefore, we can extend the angular velocity equality of Eq. (1.188) to include the quaternion rates:

$$\dot{a}_i^\alpha = a_i^\beta \bar{W}_\beta^\alpha = a_\rho^\alpha W_i^\rho = B_i^{\alpha\rho}\dot{\theta}_\rho = 2Q_i^{\alpha\rho}\dot{q}_\rho = 2S_i^\beta \dot{T}_\beta^\alpha \tag{1.212}$$

This equation relates all forms of angular velocity measurements, i.e., the body axis (coordinate axis) rates; $\bar{\omega}_i$, the relative inertial axis rates; ω_i, the Euler angle rates; $\dot{\theta}_\rho$, and the quaternion rates $\dot{q}_\rho$.

As in Sec. 1.5.5, which determines the relationships between the body axis rotation rates and the Euler angle rates, the relationships between the body axis rotation rates and the quaternion rates are needed in many problems of rotational dynamics. By using Eq. (1.212), we can write

$$a_i^\beta \bar{W}_\beta^\alpha = S_i^\gamma T_\gamma^\beta \bar{W}_\beta^\alpha = 2S_i^\gamma \dot{T}_\gamma^\alpha$$

and using the last two expressions of this equality and subtracting,

$$S_i^\gamma T_\gamma^\beta \bar{W}_\beta^\alpha - 2S_i^\gamma \dot{T}_\gamma^\alpha = 0$$

By factoring out S_i^γ and recalling Eq. (1.118), i.e., $(T)^T(T) = (I)$, we have

$$\bar{W}_i^\alpha = 2T_{\gamma i}\dot{T}^{\gamma\alpha} \tag{1.213}$$

Notice that γ is summed from 1 to 4, but the expression reduces to the 3 × 3, skew-symmetric matrix of body axis rotation rates and in matrix form is simply $(\bar{W}) = 2(T)^T(\dot{T})$. From Eq. (1.213), the $\bar{\omega}_i$ are

$$\begin{pmatrix} \bar{\omega}_1 \\ \bar{\omega}_2 \\ \bar{\omega}_3 \end{pmatrix} = 2\begin{pmatrix} -q_2 & +q_1 & +q_4 & -q_3 \\ -q_3 & -q_4 & +q_1 & +q_2 \\ -q_4 & +q_3 & -q_2 & +q_1 \end{pmatrix}\begin{pmatrix} \dot{q}_1 \\ \dot{q}_2 \\ \dot{q}_3 \\ \dot{q}_4 \end{pmatrix} \tag{1.214}$$

The solution for the time derivatives of the quaternion as functions of the quaternion and the body axis rotation rates is also used in aerospace problems describing rotational dynamics. The following describes a very interesting application of the derivative of the quaternion as developed by Carroll in Ref. 16. Using this quaternion algebra applied to 4 × 4 matrix operations, we can start with $\mathcal{A}_i^j$ given in Eq. (1.120). Analogous to Eq. (1.173), the 4 × 4 matrix relationship is

$$(\mathcal{A})^T\left(\frac{d\mathcal{A}}{dt}\right) = (\mathcal{W}) = \begin{pmatrix} 0 & \bar{\omega}_1 & \bar{\omega}_2 & \bar{\omega}_3 \\ -\bar{\omega}_1 & 0 & -\bar{\omega}_3 & \bar{\omega}_2 \\ -\bar{\omega}_2 & \bar{\omega}_3 & 0 & -\bar{\omega}_1 \\ -\bar{\omega}_3 & -\bar{\omega}_2 & \bar{\omega}_1 & 0 \end{pmatrix} \tag{1.215}$$

Using Eq. (1.121) and, for simplicity, continuing in matrix form, we can write

$$\left(\frac{\mathrm{d}\mathcal{A}}{\mathrm{d}t}\right) = (\mathcal{S})(\mathcal{T})(\bar{\mathcal{W}}) = 2(\mathcal{S})(\dot{\mathcal{T}})$$

Remembering that $(\mathcal{S})^T(\mathcal{S}) = (\mathcal{I})$, i.e., the $\mathcal{S}$ matrix is orthogonal, we can solve for $\dot{\mathcal{T}}$ as follows:

$$(\dot{\mathcal{T}}) = \frac{1}{2}(\mathcal{T})(\bar{\mathcal{W}}) \tag{1.216}$$

From the elements of the $\dot{\mathcal{T}}$ matrix, we have the required relationship:

$$\begin{pmatrix} \dot{q}_1 \\ \dot{q}_2 \\ \dot{q}_3 \\ \dot{q}_4 \end{pmatrix} = \frac{1}{2} \begin{pmatrix} 0 & -\bar{\omega}_1 & -\bar{\omega}_2 & -\bar{\omega}_3 \\ \bar{\omega}_1 & 0 & \bar{\omega}_3 & -\bar{\omega}_2 \\ \bar{\omega}_2 & -\bar{\omega}_3 & 0 & \bar{\omega}_1 \\ \bar{\omega}_3 & \bar{\omega}_2 & -\bar{\omega}_1 & 0 \end{pmatrix} \begin{pmatrix} q_1 \\ q_2 \\ q_3 \\ q_4 \end{pmatrix} \tag{1.217}$$

1.5.7 Aerospace Applications: Determining Body Axis Rotation Rates from a Moving Three-Axis Euler Coordinate Frame

We can base this example on the discussion of the analysis leading up to Eq. (1.208) in Sec. 1.5.5. Again it is instructive to select an Euler sequence that we have previously studied, namely the aeronautical standard yaw-pitch-roll sequence (Ref. 19, p. 126) as shown in Fig. 1.17. The notation used in Eq. (1.103) can be applied similarly here to describe the transformation matrix for this 3-2-1 Euler sequence:

$$(a) = Z(\Psi)Y(\Theta)X(\Phi) \tag{1.218}$$

Very carefully we can write out the R_n matrices (the single-axis rotation matrices) for Eq. (1.208) in the sequence of rotations as they occur: $R_1 = Z(\Psi)$, $R_2 = Y(\Theta)$, and $R_3 = X(\Phi)$. The single-axis rotation matrices are given in Eqs. (1.83), (1.82), and (1.81), in this sequence for this application, and Eq. (1.208) becomes

$$\begin{pmatrix} \bar{\omega}_1 \\ \bar{\omega}_2 \\ \bar{\omega}_3 \end{pmatrix} = \begin{pmatrix} p \\ q \\ r \end{pmatrix} = \begin{pmatrix} 1 & 0 & 0 \\ 0 & \cos\Phi & \sin\Phi \\ 0 & -\sin\Phi & \cos\Phi \end{pmatrix}$$

$$\times \left[\begin{pmatrix} \cos\Theta & 0 & -\sin\Theta \\ 0 & 1 & 0 \\ \sin\Theta & 0 & \cos\Theta \end{pmatrix} \begin{pmatrix} 0 \\ 0 \\ \dot{\Psi} \end{pmatrix} + \begin{pmatrix} 0 \\ \dot{\Theta} \\ 0 \end{pmatrix} \right] + \begin{pmatrix} \dot{\Phi} \\ 0 \\ 0 \end{pmatrix}$$

which reduces to

$$\begin{pmatrix} p \\ q \\ r \end{pmatrix} = \begin{pmatrix} -\sin\Theta\dot{\Psi} + \dot{\Phi} \\ \cos\Phi\dot{\Theta} + \sin\Phi\cos\Theta\dot{\Psi} \\ -\sin\Phi\dot{\Theta} + \cos\Phi\cos\Theta\dot{\Psi} \end{pmatrix} \tag{1.219}$$

the body axis rotation rates as a function of the Euler angles and their rates of change. These are the same equations as derived geometrically and resulting in Eq. (1.194).

1.6 Transformation Equations for Velocity and Acceleration

1.6.1 Velocity Transformation

The derivative of the position components from Eq. (1.169) (which may be referred to as the *forward transformation*) can be written using the derivative of the transformation matrix from Eq. (1.183):

$$v_i = a_i^\beta \bar{W}_\beta^\alpha \bar{x}_\alpha + a_i^\beta \bar{v}_\beta + V_i \tag{1.220}$$

Factoring out the transformation matrix, we can write

$$v_i = a_i^\beta \left(\bar{W}_\beta^\alpha \bar{x}_\alpha + \bar{v}_\beta\right) + V_i \tag{1.221}$$

If we define

$$\bar{\nu}_\beta = \bar{W}_\beta^\alpha \bar{x}_\alpha + \bar{v}_\beta \tag{1.222}$$

which is the motion with respect to the barred reference frame due to relative position $\bar{x}$ and relative velocity $\bar{v}$, then Eq. (1.221) becomes

$$v_i = a_i^\beta \bar{\nu}_\beta + V_i \tag{1.223}$$

Notice that the form of Eq. (1.223) is the same as Eq. (1.168), the position component transformation equation. This shows that after differentiation of Eq. (1.168), the resulting vector component equation has the same form. This demonstrates the *tensor character* of velocity vector components.

Continuing to the *reverse transformation*, the position vector in the moving frame is given by Eq. (1.15),

$$\bar{x}^i = a_\alpha^i (x^\alpha - X^\alpha) \tag{1.224}$$

For simplicity in this analysis, we can use the notation $\mathcal{X}^\alpha = x^\alpha - X^\alpha$, and so Eq. (1.224) becomes

$$\bar{x}^i = a_\alpha^i \mathcal{X}^\alpha \tag{1.225}$$

Differentiating this equation and using the angular velocity equality in Eq. (1.188), we can write the components of velocity in the moving frame:

$$\bar{v}^i = a_\rho^i W_\alpha^\rho \mathcal{X}^\alpha + a_\rho^i \mathcal{V}^\rho \tag{1.226}$$

where we let $\mathcal{V}^\rho = v^\rho - V^\rho$. Again for simplicity in the analysis and factoring out summations of a_ρ^i,

$$\bar{v}^i = a_\rho^i \left(W_\alpha^\rho \mathcal{X}^\alpha + \mathcal{V}^\rho\right) \tag{1.227}$$

In a similar manner to Eq. (1.222), we let

$$\nu^\rho = W_\alpha^\rho \mathcal{X}^\alpha + \mathcal{V}^\rho \tag{1.228}$$

and Eq. (1.227) simply becomes

$$\bar{v}^i = a^i_{\rho}\nu^{\rho} \tag{1.229}$$

which is the same form as Eq. (1.224) and demonstrates the *tensor character* of the relative velocity vector components.

In many aerospace applications where the velocity relative to the moving frame is required, the body axis rotation rates are known, and the $\bar{W}$ matrix is readily available. For this reason, the *reverse transformation* for the relative velocity is found by solving for the $\bar{v}^i$ vector components from Eq. (1.221), and we have

$$\bar{v}^i = a^i_{\rho}\mathcal{V}^{\rho} - \bar{W}^{i\alpha}\bar{x}_{\alpha} \tag{1.230}$$

or in matrix form simply $\bar{\boldsymbol{v}} = (a)^T\boldsymbol{\mathcal{V}} - (\bar{W})\bar{\boldsymbol{x}}$.

1.6.2 Acceleration Transformation

The acceleration vector components in the unbarred coordinates (the inertial frame) can be computed by differentiating Eq. (1.220), factoring sums of the transformation matrix and grouping the terms of $\bar{x}_{\alpha}$ as follows:

$$\dot{v}_i = a^{\gamma}_i\left[\left(\bar{W}^{\beta}_{\gamma}\bar{W}^{\alpha}_{\beta} + \dot{\bar{W}}^{\alpha}_{\gamma}\right)\bar{x}_{\alpha} + 2\bar{W}^{\rho}_{\gamma}\bar{v}_{\rho} + \dot{\bar{v}}_{\gamma}\right] + \dot{V}_i \tag{1.231}$$

As in the velocity transformation analysis, we can group the relative motion terms into a single vector and define

$$\bar{a}_{\gamma} = \left(\bar{W}^{\beta}_{\gamma}\bar{W}^{\alpha}_{\beta} + \dot{\bar{W}}^{\alpha}_{\gamma}\right)\bar{x}_{\alpha} + 2\bar{W}^{\rho}_{\gamma}\bar{v}_{\rho} + \dot{\bar{v}}_{\gamma} \tag{1.232}$$

and then the acceleration vector components will also display the Cartesian tensor characteristics, i.e., Eq. (1.231) simply becomes

$$\dot{v}_i = a^{\gamma}_i\bar{a}_{\gamma} + \dot{V}_i \tag{1.233}$$

Notice again that after repeated differentiation with respect to time the form of the component equations remain the same.

We can further simplify the expression for $\dot{v}_i$ by grouping the angular velocity matrices of $\bar{a}_{\gamma}$ and by defining

$$\bar{\Omega}^{\alpha}_{\gamma} = \left(\bar{W}^{\beta}_{\gamma}\bar{W}^{\alpha}_{\beta} + \dot{\bar{W}}^{\alpha}_{\gamma}\right) \tag{1.234}$$

Carrying out the indicated matrix operations of Eq. (1.234), we can write

$$\bar{\Omega}^{\alpha}_{\gamma} = \begin{pmatrix} \left(-\bar{\omega}_2^2 - \bar{\omega}_3^2\right) & (\bar{\omega}_1\bar{\omega}_2 - \dot{\bar{\omega}}_3) & (\bar{\omega}_1\bar{\omega}_3 + \dot{\bar{\omega}}_2) \\ (\bar{\omega}_1\bar{\omega}_2 + \dot{\bar{\omega}}_3) & \left(-\bar{\omega}_1^2 - \bar{\omega}_3^2\right) & (\bar{\omega}_2\bar{\omega}_3 - \dot{\bar{\omega}}_1) \\ (\bar{\omega}_1\bar{\omega}_3 - \dot{\bar{\omega}}_2) & (\bar{\omega}_2\bar{\omega}_3 + \dot{\bar{\omega}}_1) & \left(-\bar{\omega}_1^2 - \bar{\omega}_2^2\right) \end{pmatrix} \tag{1.235}$$

Equation (1.232) can be written

$$\bar{a}\gamma = \bar{\Omega}^{\alpha}_{\gamma}\bar{x}_{\alpha} + 2\bar{W}^{\rho}_{\gamma}\bar{v}_{\rho} + \dot{\bar{v}}_{\gamma} \tag{1.236}$$

The acceleration in the barred reference frame (the moving frame) is given in Eq. (1.232) or (1.236), and each term may be described as follows:

1) $\bar{W}_{\gamma}^{\beta}\bar{W}_{\beta}^{\alpha}\bar{x}_{\alpha}$, the centrifugal acceleration, due to the position of the point in the barred reference frame.

2) $\dot{\bar{W}}_{\gamma}^{\alpha}\bar{x}_{\alpha}$, has no historical name, but is the acceleration due to the position of the point in the moving frame and changing body axis rotation rates.

3) $2\bar{W}_{\gamma}^{\rho}\bar{v}_{\rho}$, the Coriolis acceleration, due to the velocity of the moving point in the barred reference frame.

4) $\dot{\bar{v}}_{\gamma}$, the acceleration of the moving point in the barred reference frame.

Likewise, as with the velocity vector components, the *reverse transformation* for the acceleration components relative to the moving system may be found by differentiating Eq. (1.226) with respect to time as follows:

$$\dot{\bar{v}}^{i} = a_{\gamma}^{i}\left[\left(W_{\rho}^{\gamma}W_{\alpha}^{\rho} + \dot{W}_{\alpha}^{\gamma}\right)\mathcal{X}^{\alpha} + 2W_{\rho}^{\gamma}\mathcal{V}^{\rho} + (\dot{v}^{\gamma} - \dot{V}^{\gamma})\right] \tag{1.237}$$

Again as in the *forward acceleration transformation* of Eq. (1.232), we can define

$$\bar{b}^{\gamma} = \left(W_{\rho}^{\gamma}W_{\alpha}^{\rho} + \dot{W}_{\alpha}^{\gamma}\right)\mathcal{X}^{\alpha} + 2W_{\rho}^{\gamma}\mathcal{V}^{\rho} + (\dot{v}^{\gamma} - \dot{V}^{\gamma}) \tag{1.238}$$

and the relative acceleration components display tensor character,

$$\dot{\bar{v}}^{i} = a_{\gamma}^{i}\bar{b}^{\gamma} \tag{1.239}$$

Likewise, as with the reverse velocity transformation in Eq. (1.230), the body axis rotation matrices are more generally available in aerospace analysis (in contrast with the inertial rate matrices). Therefore, the solution for the $\dot{\bar{v}}_i$ vector components is easily formed from Eq. (1.231) as follows:

$$\dot{\bar{v}}_{i} = a_{\alpha i}(\dot{v}^{\alpha} - \dot{V}^{\alpha}) - \left(\bar{W}_{i}^{\beta}\bar{W}_{\beta}^{\gamma} + \dot{\bar{W}}_{i}^{\gamma}\right)\bar{x}_{\gamma} - 2\bar{W}_{i}^{\rho}\bar{v}_{\rho} \tag{1.240}$$

1.6.3 Aerospace Applications: Transformations of the State Vector

The Earth-centered, inertial (ECI) position vector $\boldsymbol{x}$ and the velocity vector $\boldsymbol{v}$ are sometimes combined into a single six-vector called the *state vector*, $(x_1, x_2, x_3, v_1, v_2, v_3)$ at some time t. The time of the state vector is referred to as the epoch of the vector or simply the time the state vector is valid. This state vector representation is useful because it defines both the potential and kinetic energy states of a space vehicle at any time and can be used as the initial conditions for the analysis of the motion of the vehicle. In many applications the state vector is operated on as a single vector. For instance, we can combine Eqs. (1.168) and (1.221) into

$$\begin{pmatrix} x_1 \\ x_2 \\ x_3 \\ v_1 \\ v_2 \\ v_3 \end{pmatrix} = \begin{pmatrix} a & 0 \\ a\bar{W} & a \end{pmatrix} \begin{pmatrix} \bar{x}_1 \\ \bar{x}_2 \\ \bar{x}_3 \\ \bar{v}_1 \\ \bar{v}_2 \\ \bar{v}_3 \end{pmatrix} + \begin{pmatrix} X_1 \\ X_2 \\ X_3 \\ V_1 \\ V_2 \\ V_3 \end{pmatrix} \tag{1.241}$$

Notice in the notation here that the parentheses have been omitted from the matrix partitions that form the 6×6 transformation matrix.

Likewise, the reverse transformation is formed from Eqs. (1.224) and (1.226) as follows:

$$\begin{pmatrix} \bar{x}_1 \\ \bar{x}_2 \\ \bar{x}_3 \\ \bar{v}_1 \\ \bar{v}_2 \\ \bar{v}_3 \end{pmatrix} = \begin{pmatrix} a^T & 0 \\ -\bar{W}a^T & a^T \end{pmatrix} \begin{pmatrix} x_1 - X_1 \\ x_2 - X_2 \\ x_3 - X_3 \\ v_1 - V_1 \\ v_2 - V_2 \\ v_3 - V_3 \end{pmatrix} \tag{1.242}$$

From these two equations, we have the expected results that

$$\begin{pmatrix} a & 0 \\ a\bar{W} & a \end{pmatrix}^{-1} = \begin{pmatrix} a^T & 0 \\ -\bar{W}a^T & a^T \end{pmatrix} \tag{1.243}$$

In some aerospace applications it may be necessary or advantageous to perform the motion integrals in the Earth-centered, Earth-fixed (ECF) coordinate frame. For this example, the ECF state vector can be represented by barred coordinates $(\bar{x}_1, \bar{x}_2, \bar{x}_3, \bar{v}_1, \bar{v}_2, \bar{v}_3)$. The ECF frame is attached to the Earth and therefore rotates with respect to the ECI frame. The rotation occurs about the Earth's polar axis. The ECF $\bar{x}_1$ axis is related to the ECI frame by a single angle θ and varies at a constant rate $\bar{\omega}_e$, the *sidereal rotation rate* (the rotation rate of the Earth relative to the stars) of the Earth. At any given time, we have

$$\theta(t) = \theta(t_o) + \bar{\omega}_e(t - t_o) \tag{1.244}$$

The components of the transformation matrix a are given by a single z-axis rotation given in Eq. (1.83), and we have

$$(a) = \begin{pmatrix} \cos\theta & -\sin\theta & 0 \\ \sin\theta & \cos\theta & 0 \\ 0 & 0 & 1 \end{pmatrix} \tag{1.245}$$

The $\bar{W}$ matrix in these equations is formed from the Earth's rotation rate vector $(0, 0, \bar{\omega}_e)$ and from Eq. (1.182) and becomes

$$(\bar{W}) = \begin{pmatrix} 0 & -\bar{\omega}_e & 0 \\ \bar{\omega}_e & 0 & 0 \\ 0 & 0 & 0 \end{pmatrix} \tag{1.246}$$

The acceleration vector in the ECF frame is defined by Eq. (1.232), and since the Earth's rotation rate $\bar{\omega}_e$ is considered constant, the $\dot{\bar{W}}$ matrix is zero, and Eq. (1.232) becomes

$$\begin{pmatrix} \bar{a}_1 \\ \bar{a}_2 \\ \bar{a}_3 \end{pmatrix} = \begin{pmatrix} -\bar{\omega}_e^2 & 0 & 0 \\ 0 & -\bar{\omega}_e^2 & 0 \\ 0 & 0 & 0 \end{pmatrix} \begin{pmatrix} \bar{x}_1 \\ \bar{x}_2 \\ \bar{x}_3 \end{pmatrix} + 2 \begin{pmatrix} 0 & -\bar{\omega}_e & 0 \\ \bar{\omega}_e & 0 & \\ 0 & 0 & 0 \end{pmatrix} \begin{pmatrix} \bar{v}_1 \\ \bar{v}_2 \\ \bar{v}_3 \end{pmatrix} + \begin{pmatrix} \dot{\bar{v}}_1 \\ \dot{\bar{v}}_2 \\ \dot{\bar{v}}_3 \end{pmatrix} \tag{1.247}$$

The $\dot{\bar{v}}_i$ are the components of the acceleration vector due to the forces acting on the space vehicle's center of mass in the ECF frame. The integrals of Eq. (1.247) describe the motion of the space vehicle in the ECF frame. Because the center of the ECF frame coincides with center of the ECI frame and the ECF coordinate center is stationary in the ECI frame, we have $X_i = 0$ and $V_i = 0$ respectively, and the state vector transformation Eq. (1.241) at time t becomes

$$\begin{pmatrix} x_1 \\ x_2 \\ x_3 \\ v_1 \\ v_2 \\ v_3 \end{pmatrix} = \begin{pmatrix} \cos\theta & -\sin\theta & 0 & 0 & 0 & 0 \\ \sin\theta & \cos\theta & 0 & 0 & 0 & 0 \\ 0 & 0 & 1 & 0 & 0 & 0 \\ -\bar{\omega}_e \sin\theta & -\bar{\omega}_e \cos\theta & 0 & \cos\theta & -\sin\theta & 0 \\ \bar{\omega}_e \cos\theta & -\bar{\omega}_e \sin\theta & 0 & \sin\theta & \cos\theta & 0 \\ 0 & 0 & 0 & 0 & 0 & 1 \end{pmatrix} \begin{pmatrix} \bar{x}_1 \\ \bar{x}_2 \\ \bar{x}_3 \\ \bar{v}_1 \\ \bar{v}_2 \\ \bar{v}_3 \end{pmatrix} \tag{1.248}$$

Conclusion

The formulas and analysis presented in this chapter provide the reader with the background and knowledge which is applied directly to problems in aerospace for use in onboard GNC software or in ground-based flight simulation programs. I have related each of the analysis topics to actual sensor and navigation functions. With this information I hope that the reader can now better understand the flight vehicle GNC processes used today in aerospace. An example is understanding how sensor measurements from the onboard star tracker or from the IMU are used to determine the inertial attitude of the flight vehicle. Then this actual vehicle attitude is compared to the desired attitude from the guidance software function to form the vehicle steering quaternion. And finally, the reader is presented with how the steering quaternion can be used to compute the vehicle body axis turning rate commands that are given to the appropriate flight control effectors to fly the space vehicle along the desired flight trajectory.

In conclusion I would like to reemphasize that the scientist or aerospace engineer must find ways to understand or to visualize the coordinate transformation matrices that are involved in the analysis of each problem. For this reason, I have shown that a set of three Euler angles can be selected from one of the 12 Euler rotational sequences to best describe the transformation matrices. However, there are inherent mathematical singularities found in each of the Euler rotational sequences. I then have shown the reader that the transformation matrices can be a function of other parameters, such as the Euler four symmetric parameters, developed in 1776 or the quaternion, developed in 1843. The four parameter mathematical methods lay dormant for almost 100 years, mainly because they are cumbersome to work with and hard for the mind to visualize. In the late 1960s computing resources became available to scientists and engineers and these methods were then applied in many mathematical applications.

By simple differentiation of the Cartesian position transformation equation, we have the velocity transformation equation. With this topic I have intended to lead the reader to understand the fundamental concepts of angular velocity. By deriving the time rate of change of the transformation matrix, I hope the reader can better visualize the vector algebra forms using the Omega cross R term. Continuing with the differentiation of the velocity transformation equation, I have derived the acceleration transformation equation. The reader can then use these transformations to better understand the derivation of the equations of motion for aerospace vehicles.

2
Motion of a Point Mass in Gravitational Space

Introduction

Using Newton's laws and representing a mass as a point in space, I mathematically describe the basic concepts of energy, angular momentum, and gravitation. These fundamental equations are written in component form by applying the summation convention. The differential equations of motion are derived for a point mass in the conservative gravitational field created by the central point mass representing a very large planet. These equations form the foundation for the Two-Body problem of astrodynamics. I then lead the reader through a detailed discussion of the solution of the Two-Body problem; the mathematics and the defining equations for the Keplerian orbits. In this chapter I have grouped two similarily derived sets of relative motion differental equations: 1) Point mass motion relative to Earth-based coordinates and 2) Point mass motion relative to space-based coordinates. For the Earth-based motion, I have included some of the basic concepts of Earth modeling which are necessary to define the motion of the center of the rotating coordinate frame, or the local Earth-fixed station frame.

I begin the space-based analysis with a mathematical discussion of the body-axis relative accelerations at any point on the space vehicle. These equations define the microgravity environment which will be experienced by the equipment onboard the orbiting space vehicle.

I have defined two solution conditions to study the relative motion equations of a point mass as seen from the space-based coordinates: 1) relative motion from space vehicles in circular orbits, such as a space station, and 2) as seen from space vehicles in an elliptical orbit, such as the motion about a space vehicle in an elliptical transfer orbit. Using Hill's approximations, I show the solution to the relative motion differential equations from the vehicle in circular orbits. When this solution is made in closed form, it is called the Clohessy and Wiltshire solution, or the C-W solution. This is an important solution to the relative motion problem and is applied to many GNC rendezvous analyses by the aerospace engineer. The two-burn rendezvous solution is presented as an aerospace application example which demonstrates the typical onboard guidance software required for rendezvous. The differential equations for the relative motion of a vehicle as seen from space vehicles in elliptical orbits do not have a closed form solution.

Another aerospace application in this chapter is the defining analyses necessary to simulate the motion of an aircraft using four degrees-of-freedom (4-DOF). In this discussion I define the aerodynamic parameters, such as the airplane angle of attack, and the computations for the aerodynamic force coefficients using first order linear variations. As an important part of any motion simulation, I introduce the flight control parameters. In this 4-DOF simulation example, the elevator deflection angle, which controls the airplane lift forces and the engine throttle setting, which controls the airplane thrust force, must be controlled to accomplish the objectives of the analysis.

2.1 Point Mass: Mathematical Descriptions

2.1.1 Newton's Second Law, Mass, Work, and Energy

In Chapter 1 of the text, the geometry of space describing the dynamics of motion was discussed in the absence of both mass and gravitation. Chapter 2 deals with the association of a point mass with the geometry of space and time. The analysis presented here is fundamental and can be found in many books on mechanics. This work is presented again here to show continuity with Cartesian tensor notation and the overall theme of the text.

If a particle of mass m^α is acted upon by a force, i.e., a push or a pull that makes physical contact with the mass, and the mass is not constrained, the mass is accelerated in inertial space in the same direction as the force. The magnitude of the acceleration is directly proportional to the force and inversely proportional to the mass:

$$\dot{v}_i^\alpha = \frac{f_i^\alpha}{m^\alpha} \tag{2.1}$$

or classically written as

$$f_i^\alpha = m^\alpha \dot{v}_i^\alpha \tag{2.2}$$

Here we borrow tensor notation for both f_i^α and $\dot{v}_i^\alpha$, where the force and acceleration vector components are *not second-order terms* but simply represent the force and acceleration components of the αth mass particle. For versatility in the the notation, we will define the first-order tensors as $f_i^\alpha = f_{\alpha i}$, and likewise the acceleration components of the αth mass particle could be written $\dot{v}_i^\alpha = \dot{v}_{\alpha i}$.

Equation (2.2) is Newton's Second Law of motion from the *PRINCIPIA* published in 1686[20]. This simple mathematical concept describing a physical phenomenon sparked the imagination of mathematicians and engineers in all the years to follow, and made it possible to predict and study the behavior of masses in motion.

Newton's laws are summarized here as follows:

1) *Newton's First Law, the Law of Inertia*: A body continues in its state of rest or of uniform motion in a straight line unless it is compelled to change that state by forces acting upon it, or stated differently, when the vector sum of forces on a body are zero, the acceleration of the body is zero. Forces causing the accelerations

can be referred to as *contact forces*, i.e., forces in actual physical contact with the body.

2) *Newton's Second Law, the Law of Momentum*: The rate of change of momentum is proportional to the applied force and is in the direction in which the force acts.

3) *Newton's Third Law, the Law of Equal Reaction*: For every action (force) on a body, there is an opposed and equal reaction on that body.

Note that Eq. (2.2) could be written

$$f_i^\alpha = \frac{\mathrm{d}}{\mathrm{d}t}(m^\alpha v_i^\alpha) \tag{2.3}$$

where

$$l_i^\alpha = m^\alpha v_i^\alpha \tag{2.4}$$

is defined as the linear momentum of the mass m^α.

If the force f_i^α exists (this is the force acting on the αth mass in the ith direction, i.e., in the x, y, or z direction) and if motion occurs, then *work* is being done on the mass m^α. Suppose the force f_i^α moves the mass m^α a distance during the time interval δt, then associating Eq. (2.2) with the position coordinates at each end of the time interval, we can write

$$f_{i_1}^\alpha x_{i_1}^\alpha - f_{i_0}^\alpha x_{i_0}^\alpha = m^\alpha \dot{v}_{i_1}^\alpha x_{i_1}^\alpha - m^\alpha \dot{v}_{i_0}^\alpha x_{i_0}^\alpha \tag{2.5}$$

We can make the time interval small enough such that $f_{i_1}^\alpha = f_{i_0}^\alpha$, and therefore $\dot{v}_{i_1}^\alpha = \dot{v}_{i_0}^\alpha$, and we have

$$f_i^\alpha (x_{i_1}^\alpha - x_{i_0}^\alpha) = m^\alpha \dot{v}_i^\alpha (x_{i_1}^\alpha - x_{i_0}^\alpha) \tag{2.6}$$

This is the definition for work, and further by definition, the *energy* contained by the mass m^α has been changed by the amount δE^α, where

$$\delta E^\alpha = \sum_{i=1}^{3} \left(f_i^\alpha (x_{i_1}^\alpha - x_{i_0}^\alpha)\right) = \sum_{i=1}^{3} \left(m^\alpha \dot{v}_i^\alpha (x_{i_1}^\alpha - x_{i_0}^\alpha)\right) \tag{2.7}$$

The rate of change of energy possessed by the mass m^α is defined to be *power*, or the time rate of change of work being performed, and is formed by dividing Eq. (2.7) by δt and taking the limit as $\delta t \to 0$:

$$\frac{\mathrm{d}E^\alpha}{\mathrm{d}t} = \sum_{i=1}^{3} \left(f_i^\alpha v_i^\alpha\right) = \sum_{i=1}^{3} \left(m^\alpha \dot{v}_i^\alpha v_i^\alpha\right) \tag{2.8}$$

The quantity E^α is a measure of the instantaneous energy possessed by the mass m^α. Using vector notation, Eq. (2.8) takes the form

$$\frac{\mathrm{d}E^\alpha}{\mathrm{d}t} = \boldsymbol{f}^\alpha \cdot \boldsymbol{v}^\alpha = m^\alpha \dot{\boldsymbol{v}}^\alpha \cdot \boldsymbol{v}^\alpha \tag{2.9}$$

where $\boldsymbol{f}^\alpha \cdot \boldsymbol{v}^\alpha$ and $\dot{\boldsymbol{v}}^\alpha \cdot \boldsymbol{v}^\alpha$ are vector dot products.

2.1.2 Kinetic Energy and the Differential Energy Theorem

A quantity of considerable importance in motion dynamics can be defined by re-examining Eq. (2.8) and concentrating on the right-hand terms. Because we are studying a single mass particle at this point in the text, we can continue with the understanding that the αth notation can be dropped, i.e., m will now simply be used to refer to the mass m^{α}. Solving for E^{α} from Eq. (2.8), we have

$$\int_{E_0}^{E_1} \mathrm{d}E = \sum_{i=1}^{3}\left(\int_{t_0}^{t_1} (f_i v_i)\,\mathrm{d}t\right) = \sum_{i=1}^{3}\left(\int_{t_0}^{t_1} (\dot{v}_i v_i)\,\mathrm{d}t\right) \tag{2.10}$$

A closed-form solution is found by integrating the first term and third term. However, in the second term, the force f_i is an unknown function of time, but nevertheless, we have the equality over the interval $t_1 - t_0$:

$$E_1 - E_0 = (1/2)m\sum_{i=1}^{3} v_{i_{(1)}}^2 - (1/2)m\sum_{i=1}^{3} v_{i_{(0)}}^2 \tag{2.11}$$

The sums $v_{(n)}^2 = v_{1_{(n)}}^2 + v_{2_{(n)}}^2 + v_{3_{(n)}}^2$ are recognized as the total velocities at the beginning of the interval and at the end of the interval, respectively, and Eq. (2.11) becomes

$$E_1 - E_0 = (1/2)mv_{(1)}^2 - (1/2)mv_{(0)}^2 \tag{2.12}$$

Therefore, a measure of the instantaneous energy of the mass m can be written

$$E = (1/2)mv^2 \tag{2.13}$$

This is the definition of the *kinetic energy* of the mass m.

Combining these results with Eq. (2.9), we have

$$\frac{\mathrm{d}E}{\mathrm{d}t} = \frac{\mathrm{d}}{\mathrm{d}t}((1/2)mv^2) = f^{\beta} v_{\beta} \tag{2.14}$$

which is the *differential energy theorem* (Ref. 21, p. 22). Note the indicated summation on β gives the vector dot product shown in Eq. (2.9) or the summation in Eq. (2.8).

2.1.3 Angular Momentum

Let us examine the motion of a single particle of mass and, for this discussion, again in a space geometry in the absences of gravitation. For example, assuming no *contact forces* are acting on the particle, the resulting motion will be a straight line through space as shown in Fig. 2.1.

Using Newton's Second Law, from Eq. (2.3) with $f_i = 0$ and integrating this equation, we have

$$mv_i = \text{const} \tag{2.15}$$

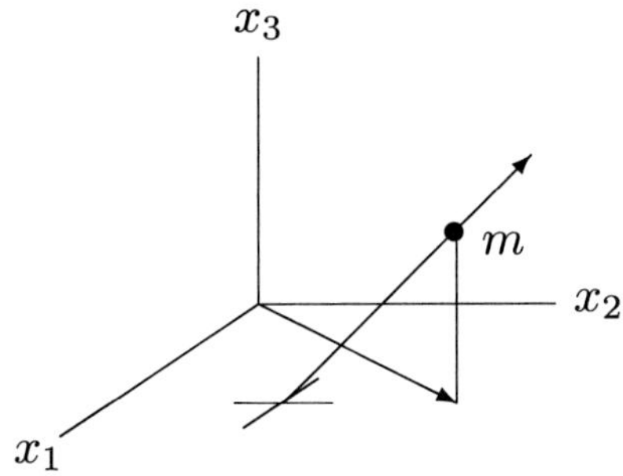

Fig. 2.1 The angular momentum of a point mass.

The linear momentum is a constant when $f_i = 0$. The projection of the position of the particle in the x_1:x_2 plane and its angular position can be measured with the following relationship:

$$\tan\theta_3 = \frac{x_2}{x_1} \tag{2.16}$$

again referring to Fig. 2.1. Angular momentum, by the very nature of its name, will be defined as a function of angular rates analogous to linear momentum being a function of linear velocity. Therefore, we can differentiate Eq. (2.16), and we have

$$\sec^2\theta_3\dot{\theta}_3 = \frac{(x_1 v_2 - x_2 v_1)}{x_1^2}$$

and

$$\dot{\theta}_3 = \frac{(x_1 v_2 - x_2 v_1)}{(x_1^2 + x_2^2)} \tag{2.17}$$

The slope of the line of track in the x_1:x_2 plane is

$$\frac{\Delta x_2}{\Delta x_1} = \frac{\mathrm{d}x_2}{\mathrm{d}x_1} = \frac{v_2}{v_1} \tag{2.18}$$

We can select any point $(x_{1(0)}, x_{2(0)}, x_{3(0)})$ along the line of track, and any other point on the line would remain in the following relationship:

$$\frac{(x_2 - x_{2(0)})}{(x_1 - x_{1(0)})} = \frac{v_2}{v_1}$$

or

$$x_1 v_2 - x_2 v_1 = v_2 x_{1(0)} - v_1 x_{2(0)} \tag{2.19}$$

The right-hand side of Eq. (2.19) is constant, and we can write Eq. (2.17) simply

$$(x_x^2 + x_2^2)\dot{\theta}_3 = x_1 v_2 - x_2 v_1 = \text{const} \tag{2.20}$$

Associating the mass particle with this equation by multiplying by m, we have the defintion of the *angular momentum* about the x_3 axis:

$$m(x_1^2 + x_2^2)\dot{\theta}_3 = m(x_1 v_2 - x_2 v_1) = L_3 \tag{2.21}$$

The angular momentum about the other axes can likewise be defined, and we have

$$\begin{aligned}
L_1 &= m(x_2 v_3 - x_3 v_2) = m(x_2^2 + x_3^2)\dot{\theta}_1 \\
L_2 &= m(x_3 v_1 - x_1 v_3) = m(x_1^2 + x_3^2)\dot{\theta}_2 \\
L_3 &= m(x_1 v_2 - x_2 v_1) = m(x_1^2 + x_2^2)\dot{\theta}_3
\end{aligned} \tag{2.22}$$

Using the left-hand equations of the preceding in matrix form,

$$\begin{pmatrix} L_1 \\ L_2 \\ L_3 \end{pmatrix} = m \begin{pmatrix} 0 & v_3 & -v_2 \\ -v_3 & 0 & v_1 \\ v_2 & -v_1 & 0 \end{pmatrix} \begin{pmatrix} x_1 \\ x_2 \\ x_3 \end{pmatrix} = m \begin{pmatrix} 0 & -x_3 & x_2 \\ x_3 & 0 & -x_1 \\ -x_2 & x_1 & 0 \end{pmatrix} \begin{pmatrix} v_1 \\ v_2 \\ v_3 \end{pmatrix} \tag{2.23}$$

and applying the summation notation, we have

$$L_i = m\Upsilon_i^{\alpha} x_\alpha = m\Lambda_i^{\alpha} v_\alpha \tag{2.24}$$

The angular momentum terms occur frequently in the analysis of vehicle motion, and these mathematical expressions are useful in many aerospace applications. In vector form Eq. (2.24) is recognized as

$$\boldsymbol{L} = m(\boldsymbol{r} \times \boldsymbol{v}) \tag{2.25}$$

Continuing with Eq. (2.3), assuming that $f_i \neq 0$, causing the mass to be accelerated and the linear momentum can no longer be constant. The resulting motion is not necessarily a straight line through space, and components of the angular momentum L_i can likewise no longer be constants. Therefore, differentiating Eq. (2.22) and using Newton's Second Law for the acceleration components, we can write the matrix form:

$$\begin{aligned}
\dot{L}_1 &= m(x_2 \dot{v}_3 - x_3 \dot{v}_2) = x_2 f_3 - x_3 f_2 = N_1 \\
\dot{L}_2 &= m(x_3 \dot{v}_1 - x_1 \dot{v}_3) = x_3 f_1 - x_1 f_3 = N_2 \\
\dot{L}_3 &= m(x_1 \dot{v}_2 - x_2 \dot{v}_1) = x_1 f_2 - x_2 f_1 = N_3
\end{aligned} \tag{2.26}$$

The N_i in this equation is the definition for the *torque* acting on the mass m. This also forms an equation analogous to Eq. (2.3) for the rate of change of the angular momentum. Simply stated, *the rate of change of the angular momentum is equal to the applied torques*. The corresponding tensor equation becomes

$$\dot{L}_i = m\dot{\Upsilon}_i^\alpha x_\alpha = m\Lambda_i^\alpha \dot{v}_\alpha = \Lambda_i^\alpha f_\alpha = N_i \tag{2.27}$$

and is also a useful mathematical relationship in the description of the motion of a particle of mass in space.

In vector notation Eq. (2.27) is found by using the vector cross product starting with Newton's Second Law, as follows:

1) Newton's Second Law in vector form,

$$\boldsymbol{f} = m\dot{\boldsymbol{v}}$$

2) and crossing with the radius vector,

$$\dot{\boldsymbol{L}} = m(\boldsymbol{r} \times \dot{\boldsymbol{v}}) = \boldsymbol{r} \times \boldsymbol{f} = \boldsymbol{N} \tag{2.28}$$

2.1.4 Aerospace Applications: The Ideal Rocket Equation

The *ideal rocket equation* is used in many guidance, navigation, and control (GNC) analysis applications. The equation provides the engineer with a relationship between the velocity gained, Δv, by the vehicle and 1) the rocket engine performance capabilities and 2) the amount of rocket propellents, both fuel and oxidizer burned. The Δv computation is made to yield estimates that are independent of vehicle attitude and are without thrust losses due to atmospheric pressure, hence the word *ideal* is used.

An important rocket engine performance parameter is the *specific impulse* (Ref. 22, pp. 540–542), which is defined as the ratio of the thrust of the engine, T, to the weight flow rate, $\dot{w}$, of the propellants being burned:

$$I_{sp} = \frac{T}{\dot{w}} \tag{2.29}$$

It will be shown in this section that the rocket engine thrust magnitude is equivalent to the rate of change of linear momentum of the exhaust gases being ejected from the rocket nozzle, $T = \dot{m}u$, where $\dot{m}$ is the propellant mass flow rate and u is the exhaust gas relative velocity. Realizing that $\dot{w} = \dot{m}g_o$, where g_o is the acceleration of gravity constant (32.174 ft/s/s), then the specific impulse is also written

$$I_{sp} = \frac{u}{g_o} \tag{2.30}$$

In this example, let the point mass m represent the instantaneous mass of rocket propelled vehicle. Writing Eq. (2.2) and placing the rocket acceleration and rocket thrust force magnitude in the velocity to be gained direction (the Δv direction),

$$m\frac{dv}{dt} = T = \dot{m}u \tag{2.31}$$

Here the problem is reduced to the ideal, one-dimensional case, i.e., in the direction of the acting thrust force. However, the thrust force direction must be accurately controlled when maximum flight performance is required. For instance, for a maximum increase in the flight vehicle's kinetic energy, *the thrust force should be directed in the instantaneous inertial velocity direction.* In any case, from Eq. (2.31), *the velocity gained will be in the thrust force direction.*

Returning to Eq. (2.31), the *contact force* generated from rocket thrust is the rate of change of linear momentum being created from the heat energy from the combustion of the rocket propellants. Equation (2.31) is a reduction (Ref. 21, pp. 173, 174) from the conservation of the linear momentum of the combined masses of rocket vehicle and the exhaust gases. Using Eq. (2.3) and realizing that the mass m is now a variable, we can write

$$\frac{\mathrm{d}(mv)}{\mathrm{d}t} = \dot{m}v + m\dot{v} = \dot{m}u + m\dot{v} = 0$$

and simply,

$$m\dot{v} = -\dot{m}u \tag{2.32}$$

Again notice in Eq. (2.32) that the vehicle acceleration is directed in the $-u$ relative velocity direction. With these details in mind, the integral of Eq. (2.32) becomes the ideal rocket equation:

$$\Delta v = g_o I_{\mathrm{sp}}\, ln\left(\frac{w_o}{w}\right) \tag{2.33}$$

Solving for the weight of propellents required for a given Δv, we have

$$\Delta w = w_o(1.0 - e^{-\varepsilon})$$

where

$$\varepsilon = \left|\frac{\Delta v}{g_o I_{\mathrm{sp}}}\right| \tag{2.34}$$

Although the word *ideal* is used here, Eqs. (2.33) and (2.34) are very good approximations when the rocket engine is fired in the vacuum of outer space. Hence, these relationships have become an integral part of the analysis used in the space vehicle's GNC software.

2.2 Point Mass and Gravitation

2.2.1 Total Energy, Gravitation, and Time

The space in the vicinity of any mass contains or is defined as the potential energy field due to gravitation. Conversely, according to Einstein (Ref. 1, pp. 142–164), the gravitation field due to mass (energy) and space can be thought of as being the same thing, i.e., space is the gravitational field.

Again, we can construct an inertial coordinate system with its origin at the center of the Earth or a large gravitating body of mass M. The large mass M is to remain stationary at this origin, which implies that $\boldsymbol{R} = 0$, and its relative velocity is $\boldsymbol{V} = 0$. The translational kinetic energy of the large mass in this inertial coordinate system is zero, i.e., $(1/2)M\,V^2 = 0$. These characteristics further define the properties of the inertial coordinate frame.

Assume that there is only one other mass particle within this space, and it will represent a space vehicle of mass m. Further assume the mass of the space vehicle is very small such that the center of mass (the center of mass will be discussed in detail in Chapter 3 of the text) of the space is unchanged due to any motion of mass m, i.e., the mass M is so large that mass m has no effect on mass M. Referring to Fig. 2.2, the total energy is the sum of the forms of energy contained by mass m:

$$E = (1/2)mv^2 - \frac{(GM)m}{r} \tag{2.35}$$

The first term is the kinetic energy as defined in Eq. (2.13), i.e., the energy of mass m due to its motion in space. The second term is the potential energy of mass m, due to gravitation and its position relative to the large mass M in space. Here, $r = |\boldsymbol{r} - \boldsymbol{R}|$, as just defined since $\boldsymbol{R} = 0$. All other forms of energy, such as heat or electrical energy, are contained and are accounted for in the mass m itself.

Referring to Eq. (1.61), the radius distance r is the length of arc in the space where the measurements x^α are taken. If we make the coordinate distances very small, we can write

$$(\delta s)^2 = g_{\alpha\beta}(\delta x)^\alpha(\delta x)^\beta \tag{2.36}$$

and remembering when the space we are in is Eucleadian, $g_{\alpha\beta} = 1$ for all $\alpha = \beta$ and zero for all $\alpha \neq \beta$. Dividing by a small variation in time squared, δt^2, and taking the limit as δt approaches zero, we have simply the velocity squared:

$$v^2 = g^{\alpha\beta}v_\alpha v_\beta = v_1^2 + v_2^2 + v_3^2 \tag{2.37}$$

By multiplying this equation by $(1/2)m$, we have the kinetic energy due to motion, the first term in Eq. (2.35),

$$E_k = (1/2)mv^2 = (1/2)mg^{\alpha\beta}v_\alpha v_\beta \tag{2.38}$$

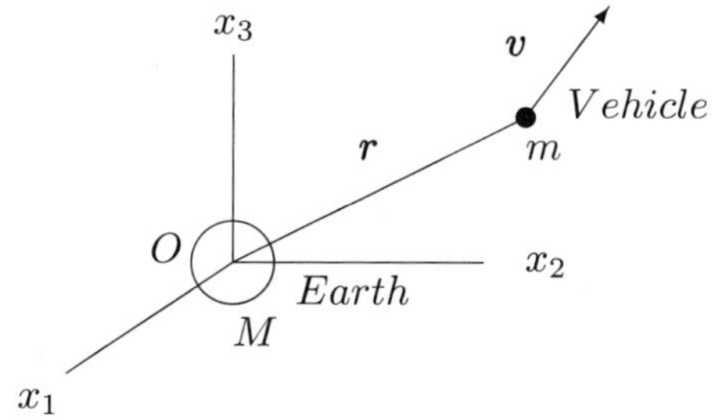

Fig. 2.2 The point mass in the gravity field of the Earth.

The second term of Eq. (2.35) is the gravitational potential, where G is the universal gravity constant (Ref. 5, p. K6) and M is the mass of the Earth or gravitating body at the center of the coordinate frame. The values for G are, in metric units:

$$G = 6.672 \times 10^{-11} m^3 kg^{-1} s^{-2}$$

and in English units:

$$G = 1.340 \times 10^{-10} (ft)^3 (slug)^{-1} s^{-2}$$

The constant μ is used for the value of GM for the large gravitating body at the center of the coordinate frame. The value for the Earth (Ref. 22, p. 5) is, in metric units:

$$\mu_e = 3.98600640 \times 10^{+14}\, m^3 s^{-2}$$

and in English units:

$$\mu_e = 1.407644875657205 \times 10^{+16} (ft)^3 s^{-2}$$

The gravitational constant for the sun (Ref. 23, p. 6) is, in metric units:

$$\mu_s = 0.1327124990 \times 10^{+21} m^3 s^{-2}$$

and in English units:

$$\mu_s = 0.4686697672 \times 10^{+22} (ft)^3 s^{-2}$$

The radius distance r from the center of the Earth is again given in Eq. (1.61) and is

$$r = (x_1^2 + x_2^2 + x_3^2)^{(1/2)} \tag{2.39}$$

Notice that as r approaches infinity, the gravitational potential in Eq. (2.35) will approach zero. Although a discussion of general relativity is beyond the scope and intent of this text, it is very interesting to point out to the reader that the term g_{44} (Ref. 1, p. 159), as a function of the gravity potential, defines to the first approximation the classical motion of the point mass. For example, if we write Eq. (1.61) using four dimensions and realizing that the units of the line element equation are length squared, we have

$$\frac{(2E)}{m} = (\delta s)^2 = (\delta x)_1^2 + (\delta x)_2^2 + (\delta x)_3^2 + g_{44}(\delta t)^2 \tag{2.40}$$

Notice when (Ref. 2, p. 124) $g_{44} = -(2\mu)/r$, Eq. (2.40) takes a form very similar to the total energy relationship given in Eq. (2.35).

Time is an ever increasing coordinate of space. All space and energy states are a function of time, and hence all mathematical descriptions of the dynamics of physical phenomenon must be related to the time coordinate of space. For example,

Table 2.1 GMT New Year's Eve calendar dates to Julian dates

GMT date	Julian date
Jan. 1, 2000	2451544.5
Jan. 1, 2001	2451910.5
Jan. 1, 2002	2452275.5
Jan. 1, 2003	2452640.5
Jan. 1, 2004	2453005.5
Jan. 1, 2005	2453371.5
Jan. 1, 2006	2453736.5

the aerospace engineer/scientist must have methods to accurately attach reference times to data measurements or to schedule mission events.

The astronomers have established the Julian date[5] as a simple method for the measurement of the time parameter (Ref. 24, Sec. 1.3). January 1, 4713 B.C., at noon, was arbitrarily selected as Julian date zero. Astronomers prefer the date to change at noon Greenwich Mean Time (GMT) so that the day numbers would not change during the nights of observations. Table 2.1 relates some GMT New Year's Eve calendar dates (Ref. 5, p. K3) to Julian dates. As an example of a reference epoch time of 18:30 hrs GMT on January 1, 2005, we have Julian date $2453371.5 + (18.5 * 3600.0)/86400.0) = 2453371.5 + 0.7708333 = 2453372.2708333$.

Notice that the number of significant figures to maintain a Julian date to fractions of seconds can exceed the computer word capacity. Usually intermediate epochs are set such as elapsed seconds from midnight on the day of the flight operation, days and seconds from January 0 of the reference year, or the use of a *modified Julian date* (Ref. 5, p. M6), where the epoch is the Julian date minus 2400000.5. The measurement of time in small increments, i.e., for very small changes in the time parameter, can be defined by using Table 2.2.

Another very interesting concept from general relativity, which can be justified experimentally, is that the *velocity of time* or rate of passage of time experienced by a mass is relative to its total energy, as given in Eq. (2.35). Hence, time passes at different rates at different places in gravitational space. For example, an object

Table 2.2 Definitions for measurement of time in small increments

Fractional part	Seconds	Cycles/second	Light travel
1	second	1 hertz	$2.9979 \times 10^{+8}$ meters
10^{-3}	millisecond	kilohertz	$2.9979 \times 10^{+5}$ meters
10^{-6}	microsecond	megahertz	299.79 meters
10^{-9}	nanosecond	gigahertz	29.979 centimeters
10^{-12}	picosecond	terahertz	0.29979 millimeters

at rest in inertial space has a total energy of simply $-\mu/r$. Using the notation $\phi = -\mu/r$, for the potential energy from Eq. (2.35), the rate of the passage of time for one second on the Earth's surface, relative to that on the surface of the sun (Ref. 2, Sec. 4.4), is predicted to be

$$\delta\tau_s = \sqrt{\frac{(1 + 2\phi_s/c^2)}{(1 + 2\phi_e/c^2)}} \tag{2.41}$$

where c is the velocity of light. The velocity of light from Ref. 23 (p. 6) is $2.997925 \times 10^{+8}$ m/s or $9.835711942257 \times 10^{+8}$ ft/s and $\phi_e = -\mu_s/R_{AU} - \mu_e/R_e$, i.e., the potential from the sun at the Earth's orbital distance plus that from the Earth on its surface. From Eq. (2.41), one second on Earth is estimated to be 0.999997888 seconds on the sun or about 2.11 μs difference in the rate of passage of time on the sun as compared to that on the Earth. Hence, *the rate of passage of time is slower in gravitational space regions with higher gravitational potentials.*

Closer to home, the Global Positioning System (GPS) satellites, which are used for worldwide navigation, depend upon very accurate distance measurements of the position of each satellite. Time errors are a major source of position errors that directly affect the navigation accuracy of the system. The GPS atomic clocks onboard each satellite must be corrected for the relativistic time dillation, which is about 0.52 ns just due to the different locations in the Earth's gravitational field. The GPS *constellation of satellites* are in circular orbits with orbital periods of 12 h and at 2.0232087×10^7 m above the Earth's surface. Hence, the light time travel corrections, as suggested in Table 2.2, must also be included for calculating accurate distance measurements.

The numerical computations for the planetary ephemerides provided by the Jet Propulsion Laboratory[25] (JPL) include relativistic effects in the n-body differential equations of motion. The JPL ephemerides match actual planetary motion with much greater accuracy than the ephemerides generated using only classical mechanics.

These are very interesting results from general relativity, especially in physics and astronomy. However, relativistic effects are generally very very small and are not necessary or even possible to be included in most aerospace engineering applications.

2.2.2 Equations of Motion for a Point Mass

The fundamental differential equations of motion for a point mass in gravitational space can now be derived. Using the assumptions and analysis of Sec. 2.2.1, we can differentate Eq. (2.35) with respect to time as follows:

$$\frac{\mathrm{d}E}{\mathrm{d}t} = mv\dot{v} + m\mu r^{-2}\dot{r} \tag{2.42}$$

where the derivative of v is

$$\dot{v} = \frac{(\dot{v}_\alpha v^\alpha)}{v} \tag{2.43}$$

and similarly the derivative of r becomes

$$\dot{r} = \frac{(x_\alpha v^\alpha)}{r} \tag{2.44}$$

The indicated summations on α will go from 1 to 3 and simply forms the vector dot products. Placing these two equations into Eq. (2.42) and factoring out the indicated dot products, we have

$$\frac{\mathrm{d}E}{\mathrm{d}t} = \left(m\dot{v}_\alpha + m\frac{\mu}{r^3}x_\alpha\right) v^\alpha \tag{2.45}$$

When there are no contact forces acting on the mass m, the rate of change of the total energy is zero, and the resulting gravitational space about the mass M is called a *conservative field*. The gravitational field produces the resulting gravitation acceleration without the gain or loss of energy in the system. However, for the more general case, using the differential energy theorem of Eq. (2.14) and demanding that the rate of change of energy in Eq. (2.45) comes from contact forces only (Ref. 26, pp. 16, 17), we can write

$$f_\alpha v^\alpha = \left(m\dot{v}_\alpha + m\frac{\mu}{r^3}x_\alpha\right) v^\alpha$$

and equating the dot product components, we have

$$f_i = m\dot{v}_i + m\frac{\mu}{r^3}x_i \tag{2.46}$$

Hence, based on the simple fundamental concept of the conservation of energy, Eq. (2.46) confirms 1) Newton's Second Law of motion and 2) Newton's Law of Gravitation[20]: *Every pair of bodies in the universe are attracted by a force which is directly proportional to their masses and inversely proportional to the square of the distance between them.*

Equation (2.46) becomes the basic equation of motion for many problems involving aircraft and spacecraft motion. In vector form Eq. (2.46) can be written

$$\boldsymbol{f} = m\dot{\boldsymbol{v}} + m\frac{\mu}{r^3}\boldsymbol{r} \tag{2.47}$$

Equations of motion, such as Eq. (2.46), can also be derived using *Lagrange's equations* (Ref. 21, Chap. 9). However, for analyses based on fundamental Cartesian inertial coordinates, as presented in this section, the derivations are easily understood and adequate for engineering applications. Likewise, nonconservative aerodynamics forces and other forces acting on an aerospace vehicle can easily be included in the equations of motion via the differential energy theorem, as given in Eq. (2.14).

In this chapter of the text, we will refer to gravitational space due to a very large single mass point only. Imagine that the mass point forms wavelike spherical surfaces of equal potential with ever increasing radii. This shape of space around a single mass point is referred to as *a spherical gravity field*. The *gravitational wave front* will move with the motion of the mass point, and its presence is propagated into space at the speed of light. In Chapter 3 we will discuss the gravitational fields resulting from a system of mass points. These gravitational fields can be considered spherical only at relatively great distances from their centers of mass.

2.2.3 Keplerian Orbit

The motion of a space vehicle in the space near a large central body is known as *the two-body problem of astrodynamics* and is studied and approximated using Eq. (2.46). The two-body problem is defined using the Cartesian inertial coordinate system as shown in Fig. 2.2. If we assume no contact forces are acting on the space vehicle, Eq. (2.47) in vector form can be written

$$\dot{\boldsymbol{v}} = -\frac{\mu}{r^3}\boldsymbol{r} \tag{2.48}$$

and becomes the differential equation of motion for a space vehicle about an *ideal planet* with the gravity constant μ or GM. The word *ideal* is used here because the gravity potential does not include the *real* nonspherical effects of the irregular mass distribution of a planet or the *real* contact forces that may be acting on a space vehicle. The equations of motion for flight vehicle simulations will be discussed in Chapter 4. Nevertheless, Eq. (2.48) and its solution are fundamentally important to the understanding of orbital mechanics and space vehicle motion.

The vehicle's total energy per unit mass from Eq. (2.35) becomes

$$\frac{E}{m} = (1/2)v^2 - \frac{\mu}{r} \tag{2.49}$$

When an initial epoch and the vehicle state vector are known, i.e., when we have $\vec{r}_o$, $\vec{v}_o$ at t_o, the starting boundary condition for Eq. (2.48) is defined and the total energy of the vehicle is determined. Because there are no contact forces acting, the total energy will be constant. Based on these definitions, the two-body problem, as given in Eq. (2.48) has a closed-form solution and defines six orbital parameters known as the *Keplerian orbital elements* (Ref. 27, p. 49), in honor of Johannes Kepler (1571–1630) (Ref. 27, p. 31). Kepler was one of the earliest contributors to the science of astrodynamics, publishing three laws in 1609 and 1618. Kepler's Laws are briefly summarized as follows:

1) The orbits of the planets are ellipses with the sun at one focus.

2) The vector connecting the sun and planets sweeps out equal areas in equal time.

3) The square of the orbital periods of the planets are proportional to the cubes of their semimajor axes.

Using the analysis based on these laws, the Keplerian orbital elements are shown to be constants and are used today to approximate the orbital motion of a vehicle in space. The elements are defined here as follows:

a = the semimajor axis of the resulting orbit
e = the orbit eccentricity
i = the orbit inclination from the inertial equatorial plane

Ω = the right ascension of the ascending node of the orbit plane
ω = the argument of pericenter
t_p = the time of pericenter passage

There are many very good derivations of the Keplerian orbital elements in the literature of astrodynamics, such as in Ref. 27 (pp. 24–55). All of these important analyses and derivations need not be reproduced in detail here; however, the application of the Keplerian elements (Ref. 8, Chap. 2) with supporting relationships will be discussed.

It is very interesting to point out that the angular momentum vector alone defines the orientation of the vehicle's orbit in the inertial reference frame, i.e., the Keplerian elements, i and Ω. Using the vector cross product on Eq. (2.48),

$$\boldsymbol{r} \times \dot{\boldsymbol{v}} = \frac{\mathrm{d}}{\mathrm{d}t}(\boldsymbol{r} \times \boldsymbol{v}) = \boldsymbol{r} \times \left(-\frac{\mu}{r^3}\boldsymbol{r}\right) = 0$$

and when performing the indicated integral, we have simply the angular momentum,

$$\boldsymbol{r} \times \boldsymbol{v} = \boldsymbol{L} = \text{a constant vector} \tag{2.50}$$

Notice that the cross product for the angular momentum vector is also defined in accordance with the right-hand rule. The components of the angular momentum vector can be transformed into the orbital coordinate frame as follows:

$$\bar{L}^i = a^i_\alpha L^\alpha \tag{2.51}$$

where the a matrix is formed from the classical Euler angle set as discussed in Sec. 1.3.1 and given in Appendix A.11. The magnitude of the angular momentum vector is invariant under orthogonal transformations.

The angle between the inertial x_3 axis and the angular momentum vector defines the orbital inclination, which is simply

$$\tan i = \frac{\sqrt{L_1^2 + L_2^2}}{L_3} \tag{2.52}$$

where L_1, L_2, and L_3 are the components of the angular momentum vector in Eq. (2.50). The orbital inclination *vector* points in the same direction as the angular momentum vector for the orbit. The orbital inclination is shown by the angle θ_2 in Fig. 1.5, where the inclination vector is coincident with the $\bar{x}_3$ axis.

From Eq. (2.52), notice that the orbit inclination i can be greater than 90 deg. When the orbit inclination is less than 90 deg, the orbiting object revolves counterclockwise (as viewed form the positive inertial x_3 axis) in its orbit around the central mass. This orbital motion is called *direct motion*. On the other hand, when the orbital inclination is greater than 90 deg, the orbiting object revolves clockwise about the central gravitating body, and the motion is called *retrograde motion*.

The right ascension of the ascending node, Ω, is defined by the line of intersection of the orbit plane and the equatorial plane of the planet. The ascending node angle is shown by the angle θ_1 in Fig. 1.4 and is discussed in Sec. 1.3.1. This line of intersection can be defined using the following vector cross product, which forms the vector $\boldsymbol{n}$, having vector components only in the equatorial plane:

$$\boldsymbol{n} = \boldsymbol{k} \times \boldsymbol{L} = n_1\boldsymbol{i} + n_2\boldsymbol{j} + (0)\boldsymbol{k}$$

and

$$\tan\Omega = \frac{n_2}{n_1} \tag{2.53}$$

There are four types of orbits that result from the Keplerian solution (Ref. 24, pp. 72, 73). The resulting orbits form conic sections in the orbital plane and are defined by the *eccentricity of the orbit*:

1) $e = 0$, *circular* orbits;
2) $0 < e < 1$, *elliptical* orbits;
3) $e = 1$, *parabolic* orbits; and
4) $1 < e < \infty$, *hyperbolic* orbits.

Keeping these orbital curve shapes in mind, we continue by selecting the y_1 axis of the orbital coordinate frame to point to the pericenter of the orbit in inertial space. The pericenter is the point in an orbit that is nearest to the center of mass of the gravitating body. This will define the *argument of pericenter* angle ω, which will be discussed in detail in the following paragraphs. Figure 2.3, which represents a space vehicle's orbit in its orbital plane, will help to clarify and define the orbital parameters used in the analysis.

From Fig. 2.3 we can write the position and velocity vectors of the space vehicle in orbital coordinates as follows:

$$\boldsymbol{r} = r\cos\nu\boldsymbol{i}' + r\sin\nu\boldsymbol{j}' + (0)\boldsymbol{k}'$$

and

$$\begin{aligned}\boldsymbol{v} &= (-v\cos\gamma\,\sin\nu + v\sin\gamma\,\cos\nu)\,\boldsymbol{i}' \\ &\quad + (v\cos\gamma\,\cos\nu + v\sin\gamma\,\sin\nu)\boldsymbol{j}' + (0)\boldsymbol{k}'\end{aligned} \tag{2.54}$$

where $\boldsymbol{i}', \boldsymbol{j}'$, and $\boldsymbol{k}'$ are the unit vectors along the orbital coordinate axes.

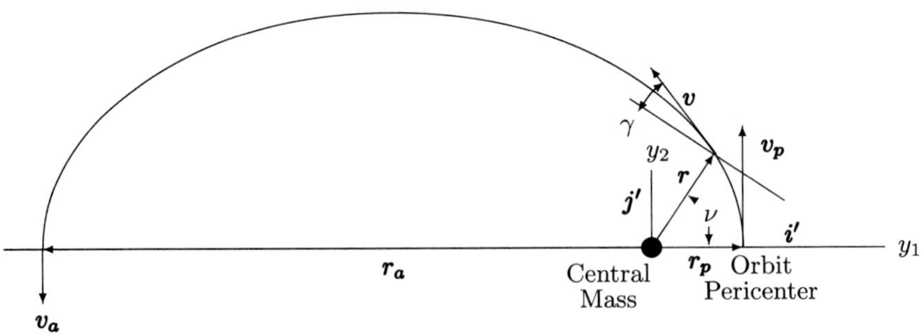

Fig. 2.3 The defining parameters in the orbit plane.

The flight path angle γ, as shown in Fig. 2.3, is the angle that the velocity vector makes with the local horizontal plane (a plane perpendicular to the radius vector $\boldsymbol{r}$) and can also be determined as a function of the angle ν, the *true anomaly* and the orbital eccentricity e, as follows:

$$\tan\gamma = \frac{e\sin\nu}{1 + e\cos\nu} \tag{2.55}$$

From the conservation of the orbital angular momentum from Eq. (2.50) and Eq. (2.51) in orbital coordinates, we have

$$v_a = v_p\left(\frac{r_p}{r_a}\right) \tag{2.56}$$

From the conservation of the total orbital energy, Eq. (2.49) can be written at the perigee and the apogee points in the orbit as follows:

$$\frac{E}{m} = (1/2)v_p^2 - \frac{\mu}{r_p}$$

and

$$\frac{E}{m} = (1/2)v_a^2 - \frac{\mu}{r_a} \tag{2.57}$$

respectively. Realizing from Fig. 2.3 that $r_a + r_p = 2a$, where a is the semimajor axis of the orbit, and using Eqs. (2.56) and (2.57), we find that the total orbital energy is only a function of the orbit semimajor axis and is given by

$$\frac{E}{m} = -\frac{\mu}{2a} \tag{2.58}$$

Using this equation in Eq. (2.49), we have the *vis-viva relationship* (Ref. 24, p. 96), which relates orbital speed to any radius distance along the orbital path:

$$v^2 = \mu\left(\frac{2}{r} - \frac{1}{a}\right) \tag{2.59}$$

By examining each type of the orbit curves using the vis-viva relationship alone, many important orbital characteristic can be better understood. Using the orbit eccentricity, as discussed in the preceding paragraphs, we have the following analysis:

1) The circular orbit, $e = 0$: We have the radius distance r, which is always equal to the semimajor axis a, along the orbit path; and therefore from Eq. (2.59), we have

$$v_c = \sqrt{\frac{\mu}{a}} \tag{2.60}$$

This is the speed required by the space vehicle to attain a circular orbit of radius a, or at an altitude of $h = a - R_e$, where R_e is the radius of the surface of the Earth. This speed is called *circular velocity*. Furthermore, the flight path angle is always zero, as well as $r_a = r_p$.

2) The elliptical orbit, $0 < e < 1$: Equation (2.59) gives the speed along the orbital path. The space vehicle's speed in excess of the circular speed at the pericenter is called the *circular excess velocity*, which will determine the vehicle's altitude rise to the apogee of the orbit. The flight path angle will pass through zero at the perigee and apogee points in the orbit and is positive during the first half of the orbit and negative during the second half.

3) The parabolic orbit, $e = 1$: The semimajor axis a approaches an infinite distance, and from Eq. (2.59), we have

$$v_e = \sqrt{\frac{2\mu}{r}} \tag{2.61}$$

which is called the *parabolic velocity* or the *escape velocity*. This is the speed required for a space vehicle to escape the gravitational field of the central mass altogether. Notice from the vis-viva relationship of Eq. (2.59), the space vehicle's relative speed will be zero at the infinite distance from the central mass.

4) The hyperbolic orbit, $1 < e < \infty$: As in the parabolic case, the space vehicle will escape the gravitational field of the central mass as well, but the direction of the motion is clearly defined and will approach the asymptote line of the hyperbola. The space vehicle's speed in excess of the parabolic velocity, given in Eq. (2.61), is called the *hyperbolic excess velocity*, which will define the asymptote line in the orbital plane and hence the targeted direction in inertial space to accomplish the mission of the space vehicle.

The vector to the pericenter of the orbit is a constant vector in inertial space that must also lie in the orbital plane, shown as $\boldsymbol{r}_p$ in Fig. 2.3. This vector is called the *Laplace vector*, or more appropriately referred to as the Laplace-Runge-Lenz vector, which is named for those who first derived this important relationship (Ref. 28, pp. 42, 43). At this point along the orbital path, the magnitude of the radius vector $|\boldsymbol{r}|$ will have a minimum value and the magnitude of the velocity vector $|\boldsymbol{v}|$ will have the maximum value. By computing the Laplace vector, we can solve for the angle ω, the argument of the pericenter, or the argument of perigee, as well as the orbital radius magnitude as a function of the true anomaly. The closest approach point in the orbit is called perigee when the Earth is the central body and the perihelion when the sun is the central body. The following derivation is presented here in Cartesian component form, which is the same as the derivation in vector form, as given in Ref. 27 (p. 23).

We begin this analysis process by forming the cross product of Eq. (2.48) with the angular momentum vector. We can create the *cross-product matrix* from the elements of the angular momentum vector, as shown in Eq. (2.24), as follows:

$$\mathcal{L}_i^\alpha = \begin{pmatrix} 0 & -L_3 & L_2 \\ L_3 & 0 & -L_1 \\ -L_2 & L_1 & 0 \end{pmatrix} \tag{2.62}$$

Multiplying this matrix times the vector components of Eq. (2.48), we have

$$\mathcal{L}_i^\alpha \dot{v}_\alpha = -\frac{\mu}{r^3}\mathcal{L}_i^\beta x_\beta \tag{2.63}$$

which forms the vector components of a vector in the orbital plane. This vector is also perpendicular to the instantaneous radius vector (x_1, x_2, x_3) and rotates in time as the space vehicle travels around in its orbit. Equation (2.63) describes the rate of change of a vector, which can be reduced as follows:

$$\mathcal{L}_i^\alpha \dot{v}_\alpha = \frac{\mathrm{d}}{\mathrm{d}t}\left(\mathcal{L}_i^\alpha v_\alpha\right) = -\frac{\mathrm{d}}{\mathrm{d}t}\left(\frac{\mu}{r}x_i\right)$$

and

$$\frac{\mathrm{d}}{\mathrm{d}t}\left(\mathcal{L}_i^\alpha v_\alpha + \frac{\mu}{r}x_i\right) = 0 \tag{2.64}$$

The integral of this equation is called the Laplace integral and forms the constant vector, the Laplace vector, in the orbital plane with components as follows:

$$\mathcal{P}_i = -\mathcal{L}_i^\alpha v_\alpha - \frac{\mu}{r}x_i \tag{2.65}$$

The magnitude of the Laplace vector is very large, on the order of $1.0 \times 10^{+16}(\mathrm{ft})^3/(\mathrm{s})^2$, but with considerable astonishment, it points to the pericenter of the orbit that is stationary in the inertial coordinate frame. Hence, the components of the Laplace vector are computed in inertial coordinates, but the argument of perigee angle ω is the angle between $\boldsymbol{n}$ of Eq. (2.53) and $\boldsymbol{\mathcal{P}}$ in the orbital plane. Numerically, it may be more expedient to work with the unit Laplace vector, $\boldsymbol{\mathcal{P}}' = \mathcal{P}/|\mathcal{P}|$, in the following equations. The unit Laplace vector can be rotated into the orbital plane coordinates by a two-axis rotation as follows:

$$\bar{\boldsymbol{\mathcal{P}}} = (Z(\Omega)X(i))^T\boldsymbol{\mathcal{P}}' \tag{2.66}$$

and the argument of perigee simply is given by

$$\tan\omega = \frac{\bar{\mathcal{P}}_2}{\bar{\mathcal{P}}_1} \tag{2.67}$$

From the Laplace vector given in Eq. (2.65), we have the definition of the orbital eccentricity (Ref. 4, p. 192):

$$e = \frac{|\boldsymbol{\mathcal{P}}|}{\mu}$$

which becomes

$$e^2 = 1 + \left(\frac{2|\boldsymbol{L}|^2}{\mu^2}\right)\left(\frac{E}{m}\right) \tag{2.68}$$

Also from Eq. (2.65), we have the solution for the radius magnitude for the orbital ellipse, which culminates a truly remarkable application of this analysis process:

$$r = \frac{a(1 - e^2)}{(1 + e \cos \nu)} \tag{2.69}$$

Again we find the *true anomaly*, ν, as shown in Fig. 2.3, which is now precisely defined as the instantaneous angle between the radius vector $\boldsymbol{r}$, to the space vehicle and the Laplace vector. The term

$$p = a(1 - e^2) \tag{2.70}$$

is called the parameter of the orbit.

With the definition of ω from the Laplace vector, the three Euler angles necessary to define the Keplerian-orbit-to-inertial transformation matrix are complete and make it possible to transform orbital coordinates y_i into the inertial coordinates x_i. This transformation, as discussed in Sec. 1.3.1, is shown in Appendix A.11 and is the three-axis transformation defined by

$$a = a(Z(\Omega), X(i), Z(\omega))$$

where

$$x_i = a_i^\alpha y_\alpha \tag{2.71}$$

Notice that for the Keplerian orbit the three angles Ω, i, and ω are constants; therefore, the orbit, as well as the orbital coordinate frame, is stationary in inertial space. For this reason, the barred notation is not necessary to represent the orbital coordinates y_i.

No discussion of the Keplerian orbit is complete without the following definitions of four very important angles that locate the space vehicle in the orbital plane:

1) The orbital longitude l_o: the angle from the ascending node $\boldsymbol{n}$ to the radius vector $\boldsymbol{r}$ of the space vehicle.

2) The true anomaly ν: the angle between the y_1 orbital coordinate axis and the radius vector $\boldsymbol{r}$. Using the transformation matrix given by Eq. (2.71), the position of the space vehicle in the orbital coordinates is $y^i = a_\alpha^i x^\alpha$ and

$$\tan \nu = \frac{y_2}{y_1} \tag{2.72}$$

or, if the true anomaly is known, the position in orbital coordinates is given in Eqs. (2.54):

$$\begin{aligned} y_1 &= r \cos \nu \\ y_2 &= r \sin \nu \\ y_3 &= 0 \end{aligned} \tag{2.73}$$

3) The eccentric anomaly E: the angle from the center of the orbit measured along the *orbit reference circle* (a circle of radius a, the semimajor axis, centered at the center of the orbit) from the y_1 axis to a point projected directly above the space vehicle on to the reference circle, such that

$$
\begin{aligned}
y_1 &= a(\cos E - e) \\
y_2 &= a(\sqrt{1 - e^2}) \sin E \\
y_3 &= 0
\end{aligned} \tag{2.74}
$$

The relationship between the true anomaly and the eccentric anomaly is found by equating the corresponding y components given in Eqs. (2.73) and (2.74) and solving for ν as follows:

$$
\nu = \tan^{-1}\left[\frac{\sqrt{1 - e^2}\sin E}{(\cos E - e)}\right] \tag{2.75}
$$

Likewise, solving for E, the eccentric anomaly, as a function of the true anomaly, we have

$$
E = \tan^{-1}\left[\frac{\sqrt{1 - e^2}\sin \nu}{(\cos \nu + e)}\right] \tag{2.76}
$$

Notice that the arctangent function is desired here to capture the anomalies over their entire range, i.e., from 0 to 360 deg.

4) The mean anomaly $\mathcal{M}$: the angle from the y_1 axis to an imaginary space vehicle traveling at a constant angular rate on the orbit reference circle as if the space vehicle would complete one revolution in the orbit in one orbital period, such that

$$
\mathcal{M}_t = n(t - t_p) \tag{2.77}
$$

where t_p is the time of pericenter passage and where $\mathcal{M}_p = 0$. The mean orbital motion n is given by

$$
n = \sqrt{\frac{\mu}{a^3}} \tag{2.78}
$$

Here the units for n are in radians per second. The period of the orbit then is

$$
P = \frac{2\pi}{n} \tag{2.79}
$$

and the units for the period P would be in seconds. These two equations demonstrate Kepler's Third Law.

The mean anomaly is related to the eccentric anomaly via Kepler's equation (Ref. 24, Sec. 3.5.1):

$$
\mathcal{M} = E - e\sin E \tag{2.80}
$$

Because of the transcendental nature of Kepler's equation, there is no direct solution for the eccentric anomaly as a function of a given mean anomaly. However, using the Newton-Raphson iterative procedure (Ref. 29, pp. 447, 448) for a given mean anomaly, we have

$$E_{i+1} = E_i - \frac{E_i - e \sin E_i - \mathcal{M}}{(1 - e \cos E_i)}, \quad i = 1, 2, \cdots n \tag{2.81}$$

This iterative procedure can be started using $E_1 = \mathcal{M}$, and continued until the value of $(E_{n+1} - E_n)$ approaches the desired accuracy and we have the eccentric anomaly as a function of the given mean anomaly.

2.2.4 Astrodynamics Applications: Space Vehicle Motion Using the Keplerian Orbital Elements

The complete transformation from a space vehicle's inertial state vector at a given epoch t_o to the Keplerian orbital elements was presented in Sec. 2.2.3. The Keplerian orbital elements are constant and represent the solution to the differential equation of motion given in Eq. (2.48). In the following analysis it is shown that this solution is valid at any other epoch time t and hence can be used to propagate the space vehicle in its Keplerian orbit. Using the Keplerian elements, the first and second integrals of Eq. (2.48) are solved! This method of orbital propagation can be applied to many problems of astrodynamics and is summarized here using *functional array notation* and the following procedures:

1) The Keplerian orbital elements are derived from the initial state vector at epoch t_o, and we can represent the processes and analysis described in Sec. 2.2.3 as follows:

$$e_i = e_i(\boldsymbol{r}_o, \boldsymbol{v}_o) = e_i(x_{o\alpha}) \tag{2.82}$$

where the functional array notation $x_{o\alpha}$ represents the state vector and where $\alpha = 1$ to 6, as described in Sec. 1.6.3. The notation as used here may be interpreted as follows: the e_i orbital element at t_o is the e_i function of $\boldsymbol{r}_o$ and $\boldsymbol{v}_o$. For this application, the Keplerian orbital element array is defined as

$e_1 = a$, the semimajor axis
$e_2 = e$, the orbit eccentricity
$e_3 = i$, the orbital inclination
$e_4 = \Omega$, the right ascension of the ascending node
$e_5 = \omega$, the argument of perigee
$e_6 = \mathcal{M}_{t_o}$, the mean anomaly at t_o, which is given by Eq. (2.77)

The mean anomaly can be expressed as $\mathcal{M}_{t_o} = n(t_o - t_p)$, simply because the initial epoch of the Cartesian state vector need not be at the time of perigee passage.

2) The Keplerian orbital elements at t_o are propagated to time t, simply by

$$\mathcal{M}_t = \mathcal{M}_{t_o} + n(t - t_o) \tag{2.83}$$

realizing that the only orbital element of e_i that will change is e_6, the mean anomaly $\mathcal{M}_t$.

3) Using the Newton-Raphson method as described in Eq. (2.81), Kepler's equation is solved for the eccentric anomaly E_t.

4) The position of the space vehicle in orbital coordinates (y_1, y_2, y_3) is computed using Eqs. (2.74).

5) The velocity of the space vehicle in orbital coordinate is computed as follows (Ref. 24, p. 113):

$$\dot{y}_1 = -\left(\frac{\sqrt{\mu a}}{r}\right)\sin E_t$$

$$\dot{y}_2 = +\left(\frac{\sqrt{\mu a(1-e^2)}}{r}\right)\cos E_t$$

$$\dot{y}_3 = 0 \tag{2.84}$$

where r is the orbit radius distance, the same as Eq. (2.69), except derived here as a function of the eccentric anomaly (Ref. 27, p. 44), which is as follows:

$$r = a(1 - e\cos E_t) \tag{2.85}$$

6) Finally we have the propagated state in the inertial coordinate frame at time t:

$$x_i = a_i^\alpha y_\alpha$$

$$\dot{x}_i = a_i^\alpha \dot{y}_\alpha \tag{2.86}$$

The orbit-to-inertial-coordinate transformation matrix (the a_i^α) is computed using Eq. (2.71). It should be noted here that both position and velocity vectors transform in the same manner, because both coordinate frames are inertial and are not rotating with respect to one another.

The solution process described in steps 1–6 can be described in functional array notation as follows:

$$x_i - x_i(e_\alpha, t) \tag{2.87}$$

where $i = 1, 3$ for the position state and

$$v_i = v_i(e_\alpha, t) \tag{2.88}$$

where $i = 1, 3$ for the velocity state. These functional evaluations propagate the inertial position and velocity states to the new epoch time t and completes the *exact solution* of Eq. (2.48) at that new epoch time. This solution method is used extensively by aerospace engineers to approximate the motion of a space vehicle and its orbit. It must be pointed out that the differential Eq. (2.48) is an ideal representation of orbital motion. Other perturbing accelerations and forces act on the real space vehicle. The differential equations of motion for actual flight vehicles will be discussed in detail in Chapter 4.

It is sometimes very convenient to define the epoch time as part of the Keplerian state definition, forming what is called a *Keplerian element set* in a single array as follows:

$e_1 = t_o$, the epoch of the Keplerian element set
$e_2 = a$, the semimajor axis
$e_3 = e$, the orbit eccentricity
$e_4 = i$, the orbit inclination
$e_5 = \Omega$, the right ascension of the ascending node
$e_6 = \omega$, the argument of perigee
$e_7 = \mathcal{M}_{t_o}$, the mean anomaly at t_o

The Keplerian orbital element set is sometimes referred to as the *osculating orbital element set* because the Keplerian elements osculate during motion in the real orbit. The osculating elements and their relationship with mean orbital element sets will be discussed in Sec. 4.4.

2.3 Point Mass Motion Relative to Earth-Based Coordinates

2.3.1 Vehicle Equations of Motion Relative to the Earth

The fundamental equations of motion developed in Sec. 2.2.2 and given in Eq. (2.46) are in the inertial coordinate frame located at the center of the Earth or the center of the large gravitating mass. In many problems of vehicle dynamics, it is necessary to work in relative coordinates, where the resulting motion can be studied and the actual measurements are being observed. For example, an Earth-based relative coordinate system, called a *topodetic reference frame*, could be an airport, a launching pad, or a tracking station. These Earth-based relative coordinate centers are attached to the Earth and are being accelerated by their attachment to the Earth's surface. Therefore, the natural motion of the coordinate center in the gravitational field is being prevented by the contact forces that hold the coordinate frames stationary on the Earth. Hence, these coordinate frames are not *inertial coordinate frames*.

The geometry defined by Eq. (1.233) can be used to represent the relative motion dynamics between an accelerated coordinate centers and a moving point mass:

$$a_i^{\gamma}\,\dot{\bar{a}}_{\gamma} = \dot{v}_i - \dot{V}_i \tag{2.89}$$

where the $\dot{V}_i$ are the components of the inertial acceleration of the center of relative frame, i.e., the airport, launching pad, or tracking station. The inertial acceleration of a point mass is found by simply solving Eq. (2.46) for the acceleration components as follows:

$$\dot{v}_i = \frac{f_i}{m} - \frac{\mu}{r^3}x_i \tag{2.90}$$

By substituting this equation for $\dot{v}_i$ into Eq. (2.89), we have all of the inertial component terms on the right-hand side of the equation as follows:

$$a_i^{\gamma}\,\dot{\bar{a}}_{\gamma} = \frac{f_i}{m} - \frac{\mu}{r^3}x_i - \dot{V}_i \tag{2.91}$$

where $\bar{a}_i$ is given by Eq. (1.232). At this point in the reduction to station relative coordinates, it is instructive to develop an alternative mathematical relationship for the gravitational acceleration term.

Using functional notation, the inertial gravitational acceleration at the station coordinate center and on the point mass not at the station coordinate center, respectively, can be written

$$G(X_i) = -\frac{\mu}{R^3}X_i$$

and

$$G(x_i) = -\frac{\mu}{r^3}x_i \tag{2.92}$$

where R is the radius distance from the center of the Earth to the station coordinate center, and r is the radius to the point mass. When the difference between these two accelerations is considered to be a variation due to a small displacement of the point mass from the station coordinate center, we have

$$\delta G(X_i) = G(x_i) - G(X_i)$$

and using the definition for the partial derivative and remembering that r is a function of x_1, x_2, and x_3, we can define a partial derivative matrix as follows:

$$P_{ij} = \lim_{(x_j - X_j)\to 0}\left[\frac{G(x_i) - G(X_i)}{(x_j - X_j)}\right] \approx \frac{\delta G(X_i)}{a_j^\alpha \bar{x}_\alpha} \tag{2.93}$$

Using this relationship, we can approximate the inertial gravitational acceleration term in Eq. (2.91) as follows:

$$\frac{\mu}{r^3}x_i = G(x_i) \approx -P_i^\beta a_\beta^\alpha \bar{x}_\alpha + \frac{\mu}{R^3}X_i \tag{2.94}$$

Substituting this relationship into Eq. (2.91), solving for the $\bar{a}_i$, and collecting the station relative terms on the left-hand side, we have

$$\bar{a}_i - \frac{\bar{f}_i}{m} - a_{\alpha i}P^{\alpha\beta}a_\beta^\rho \bar{x}_\rho = a_{\alpha i}\left(-\dot{V}^\alpha - \frac{\mu}{R^3}X^\alpha\right) \tag{2.95}$$

The matrix multiplier of the $\bar{x}_\rho$ components, the third term on the right, can be written in matrix notation as $(a)^T(P)(a)$ and is recognized as the similarity transformation (Ref. 30, pp. 317, 318), which transforms the P matrix into the station coordinate frame, i.e., $(\bar{P}) = (a)^T(P)(a)$. Equation (2.95) then becomes

$$\bar{a}_i - \frac{\bar{f}_i}{m} - \bar{P}_i^\alpha \bar{x}_\alpha = a_{\alpha i}\left(-\dot{V}^\alpha - \frac{\mu}{R^3}X^\alpha\right) \tag{2.96}$$

The partial derivative matrix $\bar{P}$ can also be computed directly by transforming the second of Eq. (2.92) into station coordinates using the fundamental position transformation in Eq. (1.168) with the following analysis:

$$G(\bar{x}) = -\mu r^{-3}a_i^\alpha \bar{x}_\alpha - \mu r^{-3}X_i$$

and by transforming the X_i components, which are the inertial vector components to the center of the Earth, into the station relative frame with $(a)^T X$, we have

$$\bar{G}(\bar{x}) = -\mu r^{-3} \begin{pmatrix} \bar{x}_1 + \bar{R}_1 \\ \bar{x}_2 + \bar{R}_2 \\ \bar{x}_3 + \bar{R}_3 \end{pmatrix} \tag{2.97}$$

where the $\bar{R}_i$ are the *components of the radius vector to the station* in station relative coordinates. In the Sec. 2.4.1, we will show that the transformation matrix a can be defined so that the space-based station relative positive $\bar{x}_3$ axis is downward toward the center of the Earth. By remembering that r is also a function of the $\bar{x}_i$ components, we can write

$$\frac{\partial r}{\partial \bar{x}_i} = \left(\frac{1}{r}\right)(\bar{x}_i + \bar{R}_i)$$

Using this relationship, the $\bar{P}$ matrix is formed by differentiating Eq. (2.97). The results are presented here for future reference:

$$\bar{P} = \frac{\partial \bar{G}}{\partial \bar{x}} = -\mu r^{-3} \begin{pmatrix} \left[1 - 3\left(\frac{(\bar{x}_1+\bar{R}_1)^2}{r^2}\right)\right] & -3\left(\frac{(\bar{x}_1+\bar{R}_1)(\bar{x}_2+\bar{R}_2)}{r^2}\right) & -3\left(\frac{(\bar{x}_1+\bar{R}_1)(\bar{x}_3+\bar{R}_3)}{r^2}\right) \\ -3\left(\frac{(\bar{x}_2+\bar{R}_2)(\bar{x}_1+\bar{R}_1)}{r^2}\right) & \left[1 - 3\left(\frac{(\bar{x}_2+\bar{R}_2)^2}{r^2}\right)\right] & -3\left(\frac{(\bar{x}_2+\bar{R}_2)(\bar{x}_3+\bar{R}_3)}{r^2}\right) \\ -3\left(\frac{(\bar{x}_3+\bar{R}_3)(\bar{x}_1+\bar{R}_1)}{r^2}\right) & -3\left(\frac{(\bar{x}_3+\bar{R}_3)(\bar{x}_2+\bar{R}_2)}{r^2}\right) & \left[1 - 3\left(\frac{(\bar{x}_3+\bar{R}_3)^2}{r^2}\right)\right] \end{pmatrix} \tag{2.98}$$

In summary, the $\bar{P}$ matrix yields the small variations in the gravitational acceleration at the point $(\bar{x}_1, \bar{x}_2, \bar{x}_3)$ in station coordinates, where the station coordinate center is located at $(\bar{R}_1, \bar{R}_2, \bar{R}_3)$, which is also given in station coordinates.

2.3.2 Earth Modeling for the Relative Motion Equations

To describe the motion of a mass relative to an Earth-based coordinate frame, we define the topodetic or station coordinate frame, which is a Cartesian frame with its coordinate axes pointing in the north, east, and vertically down directions. Precise definitions for the coordinate center location and its motion are required to determine the topodetic-to-inertial-coordinate transformation matrix, the velocity and acceleration vectors in the topodetic or the station frame. By referring to Fig. 2.4, we can identify the parameters necessary to determine the angular motion of the meridian of the station, airport, or launch pad coordinate center as it rotates on the surface of the Earth. The angle θ is given by

$$\theta = GHA + \lambda \tag{2.99}$$

GHA is the *Greenwich hour angle* and is the angle between the vector pointing to the vernal equinox (defining the inertial x_1 axis) and the *prime meridian*, which is longitude zero and is defined as the meridian of the astronomical observatory

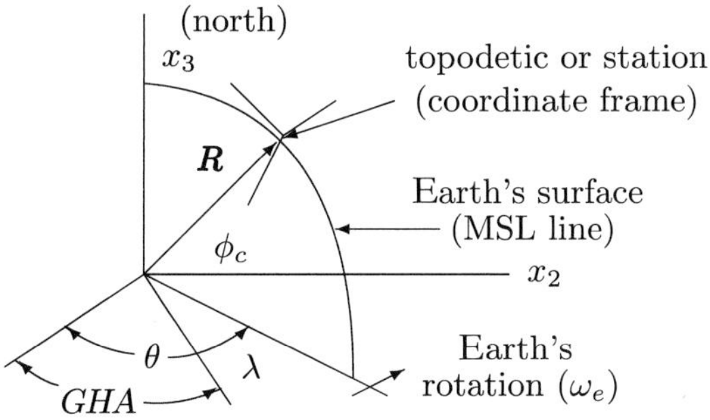

Fig. 2.4 The Earth model and the topodetic coordinate frame.

in Greenwich, England. However, the *GHA* is continually in motion due to the Earth's rotation. The angle θ is a function of time and becomes

$$\theta = GHA_o + \omega_e t + \lambda \tag{2.100}$$

where GHA_o is the right ascension of the Greenwich meridian at GMT midnight (i.e., when $t = 0$) on the day when the motion is being studied and is given in Ref. 5 (B8, Universal and Sidereal Times); ω_e is the inerital rotation rate of the Earth (the sidereal rotation rate) and is equal to $0.7292115854918357 \times 10^{-4}$ rad/s (Ref. 23, p. 2); and λ is the east longitude of the station coordinate center.

To continue the definitions for the motion of the topodetic coordinate frame, it is necessary to define a model representing the Earth's surface: the *Earth model*. The Earth is not a sphere but is approximated by an ellipsoid of revolution (Ref. 5, pp. K11, K12), and its surface is defined to be parallel to mean sea level (MSL) or the surface of water at rest at the site location. This surface is everywhere perpendicular to the acceleration vector felt by a stationary mass at the site location. The angle between the vertically downward direction or the direction of the *plumb bob line* and the Earth's axis of rotation define the co-geodetic latitude. Because of the total acceleration felt by a stationary mass at the station, the geodetic line (plumb bob line) will not pass through the center of mass of the Earth, except for points near the equator.

Figure 2.5 models the Earth's surface in the plane of the longitude of the station. Notice that in Fig. 2.5, as in Fig. 2.4, a north-east-down coordinate system is defined for the moving station frame.

The flattening or oblateness of the Earth's surface is defined by

$$f_e = \frac{(R_e - R_p)}{R_e} \tag{2.101}$$

where R_e is the equatorial radius (Ref. 31, pp. 171, 172), and R_p is the polar radius as shown in Fig. 2.5. The eccentricity of the elliptical surface is defined by

$$e = \frac{\sqrt{R_e^2 - R_p^2}}{R_e} \tag{2.102}$$

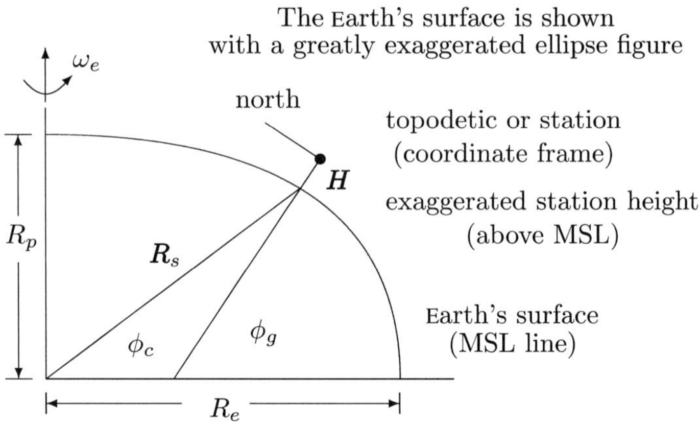

Fig. 2.5 A longitude section of the Earth model.

and

$$e^2 = 2f_e - f_e^2 \tag{2.103}$$

Using the Fischer Earth model (Ref. 32 and Ref. 23, pp. 9, 10), we have

$$f_e = \frac{1}{298.3} \tag{2.104}$$

and $e^2 = 0.6693421622965943 \times 10^{-2}$.

The relationship between the geocentric and geodetic latitudes on the Earth's surface is given by

$$\tan \phi_c = (1 - e^2) \tan \phi_g \tag{2.105}$$

where ϕ_c and ϕ_g are shown in Fig. 2.5. The magnitude of the geocentric radius vector to the surface of the Earth is

$$|\boldsymbol{R}_s| = \frac{R_e\sqrt{1 - e^2}}{\sqrt{1 - e^2 \cos^2 \phi_c}} \tag{2.106}$$

Again using the Fischer Earth model, $R_e = 20925741.47$ ft or 6378166.0 m.

It is now possible to compute the location of the station coordinate center in the Earth-centered, inertial (ECI) reference frame, which defines the X^α in Eq. (2.96):

$$\boldsymbol{R} = \begin{pmatrix} X_1 \\ X_2 \\ X_3 \end{pmatrix} = \begin{pmatrix} (R_s \cos \phi_c + H \cos \phi_g) \cos \theta \\ (R_s \cos \phi_c + H \cos \phi_g) \sin \theta \\ (R_s \sin \phi_c + H \sin \phi_g) \end{pmatrix} \tag{2.107}$$

Differentiating this equation twice with respect to time and remembering that the only time variable is θ, we can define the $\dot{V}^\alpha$ in Eq. (2.96):

$$\begin{pmatrix} \dot{V}_1 \\ \dot{V}_2 \\ \dot{V}_3 \end{pmatrix} = -\omega_e^2 \begin{pmatrix} X_1 \\ X_2 \\ 0 \end{pmatrix} \tag{2.108}$$

The inertial acceleration vector of the station coordinate center, from the right-hand side of Eq. (2.96), becomes

$$\begin{pmatrix} \mathcal{A}_1 \\ \mathcal{A}_2 \\ \mathcal{A}_3 \end{pmatrix} = \begin{pmatrix} \left(\omega_e^2 - \dfrac{\mu}{R^3}\right) X_1 \\ \left(\omega_e^2 - \dfrac{\mu}{R^3}\right) X_2 \\ -\left(\dfrac{\mu}{R^3}\right) X_3 \end{pmatrix} \tag{2.109}$$

This acceleration vector approximates the plumb bob line as the station coordinates turn with the Earth, and therefore, the angle that it makes with the equatorial plane approximates the geodetic latitude of the station as follows:

$$\tan\phi_g \cong -\frac{\mathcal{A}_3}{\sqrt{\mathcal{A}_1^2 + \mathcal{A}_2^2}} \tag{2.110}$$

where a minus sign is attached because the vertically downward plumb bob line is generally directed toward the center of the Earth, i.e., $\mathcal{A}_3$ will be negative for positive geodetic latitudes and positive for negative geodetic latitudes.

To complete the analysis for motion in the Earth-based station coordinates, which was started in Sec. 2.3.1, we can now define the transformation matrix on the right-hand side of Eq. (2.96). If we select the transformation matrix that rotates the inertial frame into the north-east-down station frame, with the $\bar{x}_3$ axis precisely aligned with the total acceleration vector of Eq. (2.109), only a single downward component of acceleration will transform into the station's rotating coordinate frame. By selecting a 3-2-1 Euler rotational sequence as described in Appendix A.10 and realizing that only two rotations are required, the transformation matrix becomes

$$a_i^\alpha = a(Z(\theta), Y((3/2)\pi - \phi_g), X(0)) \tag{2.111}$$

where θ is given by Eq. (2.100) and ϕ_g by Eq. (2.110). The right-hand side of Eq. (2.96) can be written

$$a_\alpha^i \mathcal{A}^\alpha = \begin{pmatrix} 0 \\ 0 \\ \bar{g}_s \end{pmatrix} \tag{2.112}$$

where $\bar{g}_s$ is the total acceleration felt by a mass at the Earth-based station, which is given in this analysis simply as

$$\bar{g}_s = |\boldsymbol{\mathcal{A}}| \tag{2.113}$$

Finally, using Eq. (1.232) in Eq. (2.96) and realizing that the $\dot{\bar{W}}_{\gamma}^{\alpha}$ will be zero for the Earth, we can solve for the acceleration felt by the mass m:

$$\dot{\bar{v}}_i = \frac{\bar{f}_i}{m} + \begin{pmatrix} 0 \\ 0 \\ \bar{g}_s \end{pmatrix} - (\bar{P}_i^{\alpha} + \bar{W}_i^{\beta}\bar{W}_{\beta}^{\alpha})\bar{x}_{\alpha} - 2\bar{W}_i^{\rho}\bar{v}_{\rho} \tag{2.114}$$

From the geometry shown in Fig. 2.5, the inertial rotation rate of the Earth is transformed and defines the station coordinate frame rotation rate vector:

$$\begin{pmatrix} \bar{\omega}_1 \\ \bar{\omega}_2 \\ \bar{\omega}_3 \end{pmatrix} = \begin{pmatrix} \omega_e \cos \phi_g \\ 0 \\ -\omega_e \sin \phi_g \end{pmatrix} \tag{2.115}$$

A brief analysis of the terms on the right-hand side of Eq. (2.114) shows 1) the elements of the $\bar{P}$ matrix are the magnitude of the (μ/r^3) and are about 0.1536×10^{-5} and 2) the magnitude of the $\bar{W}_i^{\beta}\bar{W}_{\beta}^{\alpha}$ terms is on the order of ω_e^2 or 0.5317×10^{-8} ft/s^2, each per foot of displacement from the station's coordinate center, respectively. These components are neglected and are generally beyond the accuracy of defining the forces causing the accelerations of the mass being studied.

The terms of $\bar{v}_{\rho}$ are the Coriolis accelerations and are on the order of $2\omega_e$ or 0.1458×10^{-3} ft/s^2 per foot per second of relative velocity. These terms cause a mass to be accelerated to the right of its velocity vector in the northern hemisphere and to the left in the southern hemisphere.

2.3.3 Aerospace Applications: Four-Degrees-of-Freedom Airframe Motion Simulation

There are many problems of aircraft performance analysis that are studied using the point mass, Earth-based relative motion equations as described by Eq. (2.114). In this analysis application, we will describe the motion of an airframe with three degrees of freedom (DOF) in Earth-based relative position coordinates and one rotational DOF about the airframe pitch axis, i.e., the $\bar{\bar{x}}_2$ airplane axis.

Notice that the Earth-based relative motion equation eliminates almost all references to the actual inertial motion. Equation (2.114) is a useful and an accurate description of airframe motion, but its application must be limited to motion relatively close to the station coordinate center. For instance, because the topodetic north-east-down coordinates are Cartesian, the height $-x_3$ measured above the north-east plane will not equal the height above the Earth as the distance from the station coordinate center becomes large due to the curvature of the Earth's surface.

The forces $\bar{f}_i$ acting on the mass m in the Earth-based reference frame in Eq. (2.114) can be determined using the linear aerodynamic force coefficients, which are used to describe the sum of the atmospheric pressure forces and the engine propulsive forces acting on the airframe. By transforming these body axis forces into the station coordinate frame, it is possible to simulate the motion of the aircraft for performance analysis or other applications. The aerodynamic force coefficients can be determined from wind-tunnel test data or by computer finite element aerodynamic modeling. In any case, the aerodynamic force coefficients are very specific for each aircraft design.

The aerodynamic force coefficients are non-dimensional coefficients (Ref. 19, p. 176) determined from the ratio of the acting force to the fluid dynamic pressure (Ref. 33, pp. 55, 56) (the total kinetic energy of the fluid passing by the aircraft) times an aircraft specific area parameter. For instance, the drag coefficient of an aircraft is

$$C_D = \frac{D}{(\bar{q}S)} \tag{2.116}$$

where D is the measured drag force magnitude; $\bar{q}$ is the dynamic pressure, which is equal to one-half the density of the air (the fluid density ρ, a function of height above MSL) times the square of the speed of the fluid passing by the aircraft, i.e., $(1/2)\rho\bar{\bar{v}}^2$; and S is the reference wing area in this example, or a specific area parameter defined by the aircraft design. The drag force acting on the aircraft is defined as *that force that is acting in the negative flight velocity direction*, as follows:

$$\boldsymbol{D} = -C_D(\bar{q}S)\left(\frac{\bar{\bar{v}}}{|\bar{\bar{v}}|}\right) \tag{2.117}$$

The total lift force is sometimes thought of as the sum of the pressure forces acting generally in the airframe vertical direction, i.e., forces in the negative airplane z-axis direction. However, more specifically the lift force L, created by an aerodynamic lifting surface, is defined as *that force acting at right angles to the flight velocity direction. The aircraft side force is simply the force acting along the instantaneous* y*-body axis of the aircraft.* The thrust force from the aircraft engine, in this simple example, is parallel to the longitudinal axis (the aircraft x axis).

The airframe axis diagram shown in Fig. 2.6 is depicted in the aircraft x-z plane and will be used to visualize and define the motion analysis parameters described in Parts 1–5 of Example 2.1.

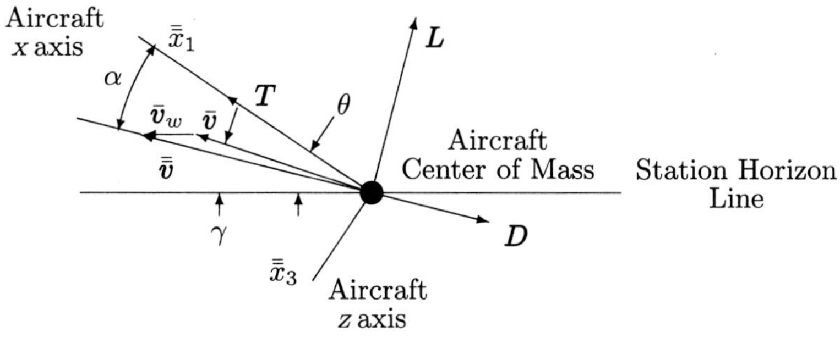

Fig. 2.6 The motion parameters for the flight vehicle.

Part 1: Computation of dynamic parameters from the aircraft state vector $(t, \bar{\boldsymbol{x}}, \bar{\boldsymbol{v}}, \theta)$ and the atmospheric wind vector $\bar{\boldsymbol{v}}_w$, both in station coordinates, as follows:

1) The aircraft pitch angle θ, as shown in Fig. 2.6, the fourth degree of freedom, which is the angle measured from the station horizontal plane to the aircraft x axis.

2) The body axis to station or airport tranformation matrix b, the airplane yaw-pitch-roll orientation with the station coordinate axes. This is the three-axis Euler transformation matrix defined by the 3-2-1 rotation sequence, where $(b) = Z(\psi)Y(\theta)X(\phi)$, as discussed in Sec. 1.5.7 and given in Appendix A.10. For this example, however, the yaw angle ψ and the roll angle ϕ will be constrained to zero.

3) The aircraft flight path angle γ, the angle measured from the horizontal plane to the aircraft's velocity vector in station relative coordinates, γ is measured positive in the upward flight direction and is given by

$$\tan \gamma = \frac{-\bar{v}_3}{\sqrt{\bar{v}_1^2 + \bar{v}_2^2}} \tag{2.118}$$

4) Aircraft aerodynamic angle of attack α, the angle measured from the aircraft x-y plane to the relative wind velocity vector. The atmospheric winds $\bar{\boldsymbol{v}}_w$ can be included at this point in the simulation. The airplane α is computed by first transforming the instantaneous station relative velocity vector into aircraft body axis coordinates with

$$\bar{\bar{v}}^i = b^i_\sigma (\bar{v}^\sigma + \bar{v}^\sigma_w) \tag{2.119}$$

and the airplane angle of attack becomes

$$\tan \alpha = \frac{\bar{\bar{v}}_3}{\sqrt{\bar{\bar{v}}_1^2 + \bar{\bar{v}}_2^2}} \tag{2.120}$$

Also notice that the airplane dynamic pressure for aerodynamic force calculations becomes $\bar{q} = (1/2)\rho(\bar{\bar{v}}_1^2 + \bar{\bar{v}}_2^2 + \bar{\bar{v}}_3^2)$.

5) The aircraft aerodynamic sideslip angle β, the angle measured from the aircraft x-z plane to the aircraft relative velocity vector, given by

$$\tan \beta = \frac{\bar{\bar{v}}_2}{\sqrt{\bar{\bar{v}}_1^2 + \bar{\bar{v}}_3^2}} \tag{2.121}$$

6) The wind axis to aircraft body axis transformation matrix b_w; the two-axis Euler tranformation matrix defined by the 3-2 rotation sequence, where $(b_w) = Z(\beta)Y(-\alpha)$:

$$(b_w) = \begin{pmatrix} \cos\beta\cos\alpha & -\sin\beta & -\cos\beta\sin\alpha \\ \sin\beta\cos\alpha & \cos\beta & -\sin\beta\sin\alpha \\ \sin\alpha & 0 & \cos\alpha \end{pmatrix} \tag{2.122}$$

Notice that the wind axis is defined such that the components of the air motion in wind axis coordinates are simply $(|\,\bar{\bar{v}}\,|, 0, 0)$. In this application, the airplane sideslip angle is usually held near zero by the *weather vaning* stability of the airplane about its z axis, and Eq. (2.122) can be reduced to

$$(b_w) \approx \begin{pmatrix} \cos\alpha & 0 & -\sin\alpha \\ 0 & 1 & 0 \\ \sin\alpha & 0 & \cos\alpha \end{pmatrix} \tag{2.123}$$

Part 2: Computations for the aerodynamic force coefficients. With these definitions it is possible to examine the aerodynamic force coefficients near an equilibrium flight condition. If we use the airplane lift coefficient, for example, and consider it to be a function of the angle of attack α, the speed of the air passing by the aircraft, $\bar{\bar{v}}$, and the dominant lift *control parameter*, the tail elevator deflection angle δ_e, we can write

$$C_L = C_L(\alpha, \bar{\bar{v}}, \delta_e) \tag{2.124}$$

The airplane lift coefficient, as used here, is based on the combined lift from the wing, body, and tail structures of the airplane. A linear variation about this flight condition, via the chain rule for differentiation, produces

$$\delta C_L \cong \frac{\partial C_L}{\partial \alpha}\Delta\alpha + \frac{\partial C_L}{\partial \bar{\bar{v}}}\Delta\bar{\bar{v}} + \frac{\partial C_L}{\partial \delta_e}\Delta\delta_e \tag{2.125}$$

which in aeronautical engineering notation is written simply

$$\delta C_L \cong C_{L_\alpha}(\alpha - \alpha_o) + C_{L_{\bar{\bar{v}}}}(\bar{\bar{v}} - \bar{\bar{v}}_o) + C_{L_{\delta_e}}(\delta_e - \delta_{e_o}) \tag{2.126}$$

where the α_o, $\bar{\bar{v}}_o$, and δ_{e_o} are the equilibrium conditions. The perturbed lift coefficient from the equilibrium lift coefficient C_{L_o} becomes $C_L = C_{L_o} + \delta C_L$. The derivatives of the aerodynamic force coefficients are called the *stability derivatives*, since their magnitudes determine the stability characteristics of the flight vehicle.

In a similar manner the drag, thrust, and side force coefficients are estimated. In this example, again remember that we are referring to an aircraft configuration and are only showing the most significant partial derivative coefficients.

The *airplane drag* includes the *induced drag* (Ref. 34, Sec. 5.14) or the drag due to the production of lift by the wing. The induced drag can accurately be estimated by the second term in the following and is presented here from Ref. 35 (pp. 90–96). The aspect ratio parameter AR for an aircraft configuration is defined as the square of the wing span divided by the reference wing area S, ϵ is the span efficiency factor or properly named the Oswald efficiency factor for Bailey Oswald, who first used this in analysis to interpret the airplane drag due to lift. For elliptical wing planforms, $\epsilon = 1.0$, and for all other wings, ϵ is less than 1.0. In this case, for subsonic airplanes, ϵ is between 0.80 and 0.95. Using these supporting definitions, the airplane drag coefficient can be written

$$C_D = C_{D_o} + \left(\frac{C_L^2}{\pi\epsilon AR}\right) + C_{D_\beta}\beta + C_{D_{\bar{\bar{v}}}}\Delta\bar{\bar{v}} + C_{D_{\delta_e}}\Delta\delta_e \tag{2.127}$$

The thrust coefficient might be estimated by the engine manufacturer over the flight envelope of the aircraft,

$$C_T = C_{T_{\delta_T}} \delta_T + C_{T_{\bar{\bar{v}}}} (\bar{\bar{v}} - \bar{\bar{v}}_o) \tag{2.128}$$

Engine performance specifications are varied and occur in many different forms depending on engine type (i.e., jet or propeller driven) and application. Engine thrust values may be given in tabular form or computed directly from engine simulation algorithms, which will be discussed in Secs. 4.2.5, 4.2.6, and 4.2.7. A simple example to estimate the engine thrust could be as follows:

$$T = \eta \left(\frac{550(HP)}{\bar{\bar{v}}} \right) \delta_T \tag{2.129}$$

where η is the propulsive efficiency, which is about 0.85 for reciprocating engine propeller installations and HP is the maximum engine horsepower.

The side force coefficient, which is based on small angle perturbations from zero, is

$$C_Y = C_{Y_\beta} \beta + C_{Y_{\delta_r}} \delta_r \tag{2.130}$$

Part 3: Defining and estimating the flight control parameters. In this 4 DOF example we have three *control parameters*:

1) $\Delta\delta_e$, the elevator deflection angle from the equilibrium condition to control the lift coefficient as well as the airplane angle of attack;

2) δ_T, the engine throttle setting, i.e., a number between 0 and 1 to represent 0–100% engine power; and

3) δ_r, the rudder deflection angle to control the side force coefficient (usually controlled to hold β, the sideslip angle, to zero).

As part of this analysis example, it is helpful to refer to the diagram shown in Fig. 1.13 of Sec. 1.4.8, which depicts the disciplines used in onboard flight systems and in motion simulation programs. Thus far, in this application, we have been working in the area of *vehicle dynamics* analysis. It is now necessary to briefly discuss the onboard GNC functions that compute the control parameters to steer the vehicle to the desired motion. The control methods that can be used in vehicle motion simulations generally are best described by two different categories, depending on the particular application:

1) Open loop: Again refer to Fig. 1.13. Here the vehicle dynamics are not fed back to active onboard sensor, navigation, guidance, and control functions. In this case these functions are not used or are not necessary for the flight of the vehicle or to simulate the motion of the vehicle for this application. More simply stated, onboard GNC functions are not being tested but the overall vehicle flight performance is being studied. For example, we could simply simulate the control of the airframe's fourth degree of freedom (the pitch angle θ) by flying a specific vehicle pitch schedule for θ and fly the aircraft at a constant power setting δ_T to evaluate the resulting performance.

2) Closed loop: This concerns vehicle motion where active onboard GNC functions are used to fly the vehicle and to simulate the motion of the vehicle. In closed-loop flight, the flight vehicle's fourth degree of freedom, i.e., the airframe's attitude, is controlled to accomplish the desired motion.

Part 4: Computations for the body axis forces. With the dynamic parameters and the control parameters defined, we can now use the aerodynamic force coefficients to yield the thrust, lift, and drag contact forces acting on the airframe. Using the wind-axis-to-airplane-body-axis transformation matrix given in Eq. (2.123) and realizing that aerodynamic drag and lift forces are defined in wind axis coordinates, i.e., in coefficient form $(-C_D, 0, -C_L)$, we have

$$
\begin{aligned}
\bar{\bar{f}}_1 &= T + (-C_D \cos\alpha + C_L \sin\alpha)\bar{q}S \\
\bar{\bar{f}}_2 &= C_Y \bar{q}S \\
\bar{\bar{f}}_3 &= -(C_D \sin\alpha + C_L \cos\alpha)\bar{q}S
\end{aligned}
\tag{2.131}
$$

Notice from Eq. (2.122) that the lift and drag forces only could contribute to the side force coefficient by the amount

$$\delta C_{Y_\beta} = -C_D \sin\beta \cos\alpha + C_L \sin\beta \sin\alpha \tag{2.132}$$

However, C_Y is better estimated with the single airplane side force coefficient as given by Eq. (2.130).

Part 5: Computing the forces in station/airport coordinates. Finally, transforming the body axis forces to station coordinates with

$$\bar{f}_i = b_i^\alpha \bar{\bar{f}}_\alpha \tag{2.133}$$

we can numerically integrate the equations of motion given by Eq. (2.114) for the airplane motion solution. When simulating flight vehicle motion on the ground, i.e., for this example vehicle motion during takeoff or landing, the ground contact force and contact rolling wheel friction forces must be included in the equations of motion. Also the vehicle motion must be constrained in the station (airport) $\bar{x}_1$-$\bar{x}_2$ plane. For this case the equations of motion from Eq. (2.114) become

$$
\begin{pmatrix} \dot{\bar{v}}_1 \\ \dot{\bar{v}}_2 \\ 0 \end{pmatrix} = \frac{1}{m} \begin{pmatrix} \bar{f}_1 + \mu_r \bar{w}_{\text{OW}} \left(\dfrac{\bar{v}_1}{|\,\bar{\boldsymbol{v}}\,|} \right) \\ \bar{f}_2 + \mu_r \bar{w}_{\text{OW}} \left(\dfrac{\bar{v}_2}{|\,\bar{\boldsymbol{v}}\,|} \right) \\ \bar{f}_3 + \bar{w}_{\text{OW}} \end{pmatrix} + \begin{pmatrix} 0 \\ 0 \\ g_s \end{pmatrix} - (\bar{P}_i^\alpha + \bar{W}_i^\beta \bar{W}_\beta^\alpha)\bar{x}_\alpha - 2\bar{W}_i^\rho \bar{v}_\rho
\tag{2.134}
$$

where $\bar{w}_{\text{OW}}$ is the *weight-on-wheels* contact force, and μ_r is the coefficient of rolling friction; μ_r can be estimated to be about 0.30 for normal runway conditions.

Here the computational procedures for each integral step might be as follows: 1) the $\bar{x}_3$ (z axis) equation of Eq. (2.134) is solved for $\bar{w}_{\text{OW}}$ and 2) this value for $\bar{w}_{\text{OW}}$ is used in the x_1 and x_2 equations of motion, and the integrals for $\dot{\bar{v}}_1$ and $\dot{\bar{v}}_2$ provide the motion of the flight vehicle in the airport $\bar{x}_1$-$\bar{x}_2$ plane. From this analysis, it is interesting to point out that the flight vehicle will lift off the runway when $\bar{w}_{\text{OW}} = 0$, thus yielding estimations for the takeoff distance performance of the flight vehicle.

Appendix C.2 details the simulation's integral list, and Appendix D.1 suggests a numerical solution method for this four-DOF airframe motion example.

2.4 Point Mass Motion Relative to Space-Based Coordinates

2.4.1 Onboard Space Vehicle Microgravity Environment

We can attach Cartesian coordinate axes to the center of mass of a space vehicle as described in Sec. 1.2.6 and as shown in Fig. 1.3. This coordinate frame is called the body axis frame. We can assume for the moment that there are no contact forces acting on the space vehicle's center of mass that coincides with the center of the body axis coordinate frame. This being the case, the ideal coordinate center will move along the *geodesic path in that space*, i.e., the trajectory followed by an object in gravitational space with unconstrained motion. The coordinate center of the inertial frame, as earlier defined, coincides with the center of mass of the gravitating planet. The body axis frame could be considered to be an inertial frame only if the attitude were held constant relative to the distant stars, i.e., with the space vehicle in the inertial-attitude-hold mode. Like the Earth-based coordinates of Sec. 2.3.1 and the LVLH coordinate frame, these frames do not qualify as inertial coordinate frames. The body axis frame is free to accelerate and to rotate in any manner, the LVLH frame is rotating in inertial space (pitching down at the instantaneous orbital rate), and the Earth-based frame is constrained to the surface of the rotating Earth. Some authors refer to the LVLH frame as a *quasi-inertial reference frame*. With these concepts in mind, we can now formalize the definition of a Cartesian inertial coordinate frame by requiring the following two conditions:

1) A Cartesian coordinate frame that is attracted to a mass that is *in motion along a geodesic path*, i.e., motion in a gravitationally conservative field with no contact forces acting on that mass. For example, using Eq. (2.35), the equations of motion for the geodesic path can be defined when $\mathrm{d}E/\mathrm{d}t = 0$.

2) The coordinated axes are *fixed in space relative to the distant stars*. This property allows the position, velocity, and acceleration transformations to reduce to the same Cartesian tensor form. This can be shown using Eqs. (1.14), (1.220), and (1.231), which reduce to $x_i = a_i^{\alpha}\bar{x}_{\alpha}$, $v_i = a_i^{\beta}\bar{v}_{\beta}$, and $\dot{v}_i = a_i^{\gamma}\dot{\bar{v}}_{\gamma}$, respectively.

For the analysis in this section, we define the inertial position to a point on the space vehicle as a function of its coordinates in the body axis frame ($\bar{\bar{x}}_i$) and the inertial position of the center of mass of the space vehicle, X_i, as follows:

$$x_i = c_i^{\alpha}\bar{\bar{x}}_{\alpha} + X_i \tag{2.135}$$

where for this case c_i^{α} are the components of the body-axis-to-inertial transformation matrix. Using this geometry, the inertial acceleration at this point on the space vehicle, from Eq. (1.231), is simply

$$\dot{v}_i = c_i^{\gamma}\left[\left(\bar{\bar{W}}_{\gamma}^{\beta}\bar{\bar{W}}_{\beta}^{\alpha} + \dot{\bar{\bar{W}}}_{\gamma}^{\alpha}\right)\bar{\bar{x}}_{\alpha} + 2\bar{\bar{W}}_{\gamma}^{\rho}\bar{\bar{v}}_{\rho} + \dot{\bar{\bar{v}}}_{\gamma}\right] + \dot{V}_i \tag{2.136}$$

The $\bar{\bar{W}}$ and the $\dot{\bar{\bar{W}}}$ matrices are the body axis rotation rate and rotation rate acceleration matrices, respectively, as given in Eq. (1.234).

We know that a small mass m at $(\bar{\bar{x}}_1, \bar{\bar{x}}_2, \bar{\bar{x}}_3)$ on the space vehicle experiences an inertial acceleration of

$$\dot{v}_i = \frac{f_i}{m} - \frac{\mu}{r^3} x_i \tag{2.137}$$

Let this small mass represent an accelerometer designed to measure the microgravity environment at its location on the space vehicle. In a similiar manner the inertial acceleration of the center of mass of the entire space vehicle is

$$\dot{V}_i = \frac{F_i}{M} - \frac{\mu}{R_3} X_i \tag{2.138}$$

where F_i is the sum of all contact forces acting at the center of mass of the space vehicle of mass M.

We now constrain the mass m to be fixed on the space vehicle, which requires that $\bar{\bar{v}}_i = 0$ and $\dot{\bar{\bar{v}}}_i = 0$ in Eq. (2.136). Using these definitions in Eq. (2.136), the inertial acceleration felt by the mass m becomes

$$\left(\frac{f_i}{m}\right) = c_i^\gamma \left[\left(\bar{\bar{W}}^\beta_\gamma \bar{\bar{W}}^\alpha_\beta + \dot{\bar{\bar{W}}}^\alpha_\gamma \right) \bar{\bar{x}}_\alpha \right] + \left(\frac{F_i}{M}\right) + \left(\frac{\mu}{r^3} x_i - \frac{\mu}{R^3} X_i \right) \tag{2.139}$$

This contact force acceleration is the microgravity acceleration required to hold the mass m in place on the space vehicle. Using the approximations leading to Eq. (2.98), the inertial acceleration at $(\bar{\bar{x}}_1, \bar{\bar{x}}_2, \bar{\bar{x}}_3)$ becomes

$$\left(\frac{f_i}{m}\right) = c_i^\gamma \left[\left(\bar{\bar{W}}^\beta_\gamma \bar{\bar{W}}^\alpha_\beta + \dot{\bar{\bar{W}}}^\alpha_\gamma \right) \bar{\bar{x}}_\alpha \right] + \left(\frac{F_i}{M}\right) + P_i^\beta c_\beta^\alpha \bar{\bar{x}}_\alpha \tag{2.140}$$

Finally, transforming this equation into body axis coordinates, we have the microgravity body axis acceleration estimate:

$$\left(\frac{\bar{\bar{f}}_i}{m}\right) = \left(\bar{\bar{W}}^\beta_i \bar{\bar{W}}^\alpha_\beta + \dot{\bar{\bar{W}}}^\alpha_i + \bar{\bar{P}}^\alpha_i \right) \bar{\bar{x}}_\alpha + \left(\frac{\bar{\bar{F}}_i}{M}\right) \tag{2.141}$$

The P matrix in Eq. (2.140) is in the inertial reference frame and is transformed into the body axis frame in Eq. (2.141) by

$$\bar{\bar{P}}^\alpha_i = a_{\gamma i} P^{\gamma\beta} a^\alpha_\beta \tag{2.142}$$

We can simplify the $\bar{\bar{P}}$ matrix representation somewhat by using the following analysis: By examination of Eq. (2.98), we realize that the $\bar{\bar{P}}$ matrix is dependent on the space vehicle's orientation relative to the inertial radius vector or to the $\bar{\bar{\boldsymbol{R}}}$ vector in the space vehicle's body axis coordinate frame. The body-axis-to-LVLH transformation matrix b can be defined using the body axis yaw-pitch-roll Euler angles relative to the LVLH coordinate axes, and we have $\bar{x}_i = b_i^\alpha \bar{\bar{x}}_\alpha$, where the $\bar{\boldsymbol{x}}_i$ are the vector components in the LVLH frame. The LVLH to inertial transformation as discussed in Sec. 1.1.4 and again in Example 1.2 of Sec. 1.3.6 is defined by $x_i = a_i^\alpha \bar{x}_\alpha$.

The c matrix in Eq. (2.135) becomes $(c) = (a)(b)$ and by examination of the operation $(\bar{\bar{P}}) = (c)^T(P)(c)$ from Eq. (2.141), we have the matrix equivalent as follows:

$$(\bar{\bar{P}}) = (b)^T(a)^T(P)(a)(b) = (b)^T[(a)^T(P)(a)](b) \tag{2.143}$$

We recognize that $(\bar{P}) = (a)^T(P)(a)$ is simply the P matrix transformed into the LVLH coordinate frame. As it will be shown in the next section, the approximation for the $\bar{P}$ matrix is much simpler, and for operational use $(\bar{\bar{P}}) = (b)^T(\bar{P})(b)$ can be better applied in Eq. (2.141).

Using the preceding analysis and Eq. (1.235) in Eq. (2.141), we again have an expression for the microgravity acceleration felt by the mass m, located at $(\bar{\bar{x}}_1, \bar{\bar{x}}_2, \bar{\bar{x}}_3)$ onboard the space vehicle:

$$\left(\frac{\bar{\bar{f}}_i}{m}\right) = (\bar{\bar{\Omega}}_i^{\alpha} + b_{\gamma i}\bar{P}^{\gamma\beta}b_{\beta}^{\alpha})\bar{\bar{x}}_\alpha + \left(\frac{\bar{\bar{F}}_i}{M}\right) \tag{2.144}$$

Notice here that even when the space vehicle is in a *quiet mode*, the aerodynamic drag force acting on the space vehicle, as discussed in Sec. 4.2.9, will cause a microgravity acceleration, i.e., $\bar{\bar{F}}_i/M$ will not be zero.

2.4.2 Vehicle Equations of Motion Relative to LVLH Coordinates

Using the analysis similiar to that developed in Sec. 2.3.1, we can derive the equations of motion for a point mass relative to an LVLH coordinate frame (see Sec. 1.1.4). This LVLH coordinate frame could be attached to a space station or another space vehicle orbiting in close proximity. We can assume for this analysis that there are no contact forces acting on the space vehicle's center of mass that represents the center of the reference LVLH coordinate frame. Starting again with Eq. (1.233) and Eq. (2.90), and using the assumption of the previous section (the motion of the reference LVLH coordinate center is along the geodesic path in gravitational space), we have

$$a_i^\gamma \ddot{\bar{a}}_\gamma = \frac{f_i}{m} - \left(\frac{\mu}{r^3}x_i - \frac{\mu}{R^3}X_i\right) \tag{2.145}$$

Using precisely the same logic presented in the development of Eqs. (2.92)–(2.93), Eq. (2.145) becomes

$$a_i^\gamma \ddot{\bar{a}}_\gamma \approx \frac{f_i}{m} + P_i^\beta a_\beta^\alpha \bar{x}_\alpha \tag{2.146}$$

Multiplying by $(a)^T$ and using $\bar{a}_i$ given by Eq. (1.232), we have the acceleration components in LVLH relative coordinates:

$$\dot{\bar{v}}_i = (\bar{P}_i^\alpha - \bar{W}_i^\beta \bar{W}_\beta^\alpha - \dot{\bar{W}}_i^\alpha)\bar{x}_\alpha - 2\bar{W}_i^\rho \bar{v}_\rho + \frac{\bar{f}_i}{m} \tag{2.147}$$

The $\bar{P}$ matrix, as given by Eq. (2.98) is in terms of the $\bar{\boldsymbol{R}}$ vector components (the LVLH vector to the center of the Earth from the LVLH coordinate center). Because the LVLH positive $\bar{x}_3$ axis is directed to the center of the Earth, the LVLH coordinate center in the inertial frame transforms as follows:

$$\bar{R}_i = a_{\alpha i} X^\alpha = \begin{pmatrix} \bar{R}_1 \\ \bar{R}_2 \\ \bar{R}_3 \end{pmatrix} = \begin{pmatrix} 0 \\ 0 \\ -R \end{pmatrix} \tag{2.148}$$

With these values for the $\bar{\boldsymbol{R}}$ vector and realizing that as the LVLH relative position components $\bar{x}_i$ become very small relative to the radius vector magnitude to the center of the Earth, we find that a very good approximation for the $\bar{P}$ matrix given by Eq. (2.98) becomes

$$\bar{P} = -\mu r^{-3} \begin{pmatrix} 1 & 0 & 0 \\ 0 & 1 & 0 \\ 0 & 0 & -2 \end{pmatrix} \tag{2.149}$$

The space-based LVLH coordinate center is generally in orbital motion about the center of the Earth and is therefore continually pitching down about the $\bar{x}_2$ axis to maintain the LVLH spacial orientation. An approximation for the LVLH coordinate frame pitch rate is the negative of the orbital rate given in Eq. (2.78), i.e., the LVLH $\bar{\omega}_i$ components are $0, -\omega_o, 0$, respectively. The orbital rate ω_o is not precisely constant as the coordinate frame goes along the orbital path, and so the $\dot{\bar{\omega}}_i$ components will be $0, \dot{\omega}_o, 0$, respectively. Using these values for $\bar{\omega}_i$ and $\dot{\bar{\omega}}_i$ and Eq. (2.149), the LVLH relative acceleration components from Eq. (2.147) become

$$\begin{aligned} \dot{\bar{v}}_1 &= ({\omega_o}^2 - \mu r^{-3})\bar{x}_1 - \dot{\omega}_o \bar{x}_3 + 2\omega_o \bar{v}_3 + (\bar{f}_1/m) \\ \dot{\bar{v}}_2 &= -\mu r^{-3}\bar{x}_2 + (\bar{f}_2/m) \\ \dot{\bar{v}}_3 &= \dot{\omega}_o \bar{x}_1 + ({\omega_o}^2 + 2\mu r^{-3})\bar{x}_3 - 2\omega_o \bar{v}_1 + (\bar{f}_3/m) \end{aligned} \tag{2.150}$$

2.4.3 *Relative Motion Equations from Space Vehicles in Circular Orbits*

In circular orbits the LVLH coordinate frame, which is attached to a reference space vehicle, will be pitching downward about the LVLH $\bar{x}_2$ axis at a nearly constant orbital rate given in Eq. (2.78). The radius distance to the center of the Earth also remains nearly constant as the vehicle orbits and is equal to the orbit semimajor axis a, and we can write

$$\omega_o^2 = \mu r^{-3} \tag{2.151}$$

Then the $\dot{\bar{\omega}}_i$ components are approximately zero, and Eq. (2.150) can be reduced to

$$\begin{aligned}\dot{\bar{v}}_1 &= 2\omega_o\bar{v}_3 + (\bar{f}_1/m)\\ \dot{\bar{v}}_2 &= -\omega_o^2\bar{x}_2 + (\bar{f}_2/m)\\ \dot{\bar{v}}_3 &= 3\omega_o^2\bar{x}_3 - 2\omega_o\bar{v}_1 + (\bar{f}_3/m)\end{aligned} \tag{2.152}$$

This approximation for the equations of motion is known as *Hill's equations*[36] and was used by Clohessy and Wiltshire[37] in 1960 to derive a closed-form solution describing the relative motion of a space vehicle as seen from another space vehicle in circular orbit. This solution, known simply as the *C-W equations*, is widely used today to predict space vehicle relative motion during rendezvous and docking operations. The C-W solution shows the cycloidal nature of the LVLH relative trajectories of two space vehicles in close proximity to each another. Figure 2.7 shows the rendezvous trajectory characteristics of a space vehicle approaching a space station in circular orbit. Orbital motion is in the plane of the space station and shown in the figure with motion from right to left on the page. This figure depicts a space vehicle with a sightly elliptical orbit having the apogee distance equal to the circular radius of the orbit of the space station. The semimajor axis and hence the orbital period of the space vehicle is less than that of the space station. The average speed of the space vehicle, as it travels from A to B (in the figure), is greater than that of the space station. The space vehicle eventually overtakes and will pass in front of the space station. The trajectory from the apogee point to the next apogee represents the relative motion during one orbital period of the space vehicle.

When using the C-W approximate solution, position and velocity errors can be detected when the separation of the two space vehicles is relatively large, say in excess of several kilometers. This is especially true today, when more accurate laser sensor devices are providing measurements to the navigation systems. Nevertheless, the C-W solution is an excellent approximation, and the errors in the relative motion are routinely removed by the GNC onboard systems during rendezvous operations. The C-W solution can be written in functional notation as follows:

$$\bar{\boldsymbol{x}}(t) = (T)\bar{\boldsymbol{x}}(t_o) \tag{2.153}$$

where the T matrix is called the *state transition matrix*. The following C-W derivation is taken from Ref. 38 and includes the *constant* acceleration components acting

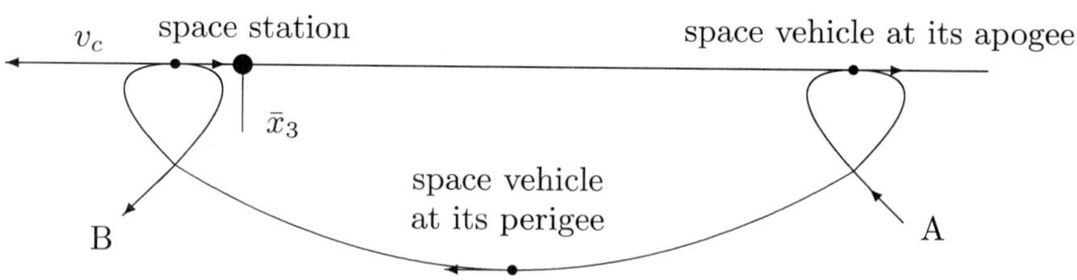

Fig. 2.7 Space vehicle relative motion as seen from the space station.

over the time interval t_o to t. The $\bar{x}(t_o)$ is the initial LVLH relative state vector and the accelerations at time t_o, a 9×1 column vector having the following form:

$$\bar{x}(t_o) = \begin{pmatrix} \bar{x}_{o1} \\ \bar{x}_{o2} \\ \bar{x}_{o3} \\ \bar{v}_{o1} \\ \bar{v}_{o2} \\ \bar{v}_{o3} \\ (\bar{f}_1/m) \\ (\bar{f}_2/m) \\ (\bar{f}_3/m) \end{pmatrix} \tag{2.154}$$

The $\bar{x}(t)$ state resulting from Eq. (2.153) is the propagated LVLH relative position and velocity at time t, a 6×1 column vector as follows:

$$\bar{x}(t) = \begin{pmatrix} \bar{x}_1 \\ \bar{x}_2 \\ \bar{x}_3 \\ \bar{v}_1 \\ \bar{v}_2 \\ \bar{v}_3 \end{pmatrix} \tag{2.155}$$

The state transition matrix T becomes a 6×9 matrix. The elements of the T matrix are presented in Appendix B. Notice that propagations either forward or backward in time are equally valid with this solution.

2.4.4 Relative Motion Equations from Space Vehicles in Elliptical Orbits

Relative motion acceleration Eqs. (2.150) *have not been solved in closed form* for a vehicle in an elliptical orbit. However, these differential equations are readily solved using numerical methods. For instance, Runge-Kutta methods have been used and have shown, as expected, to match multibody inertially integrated solutions. The relative motion trajectories from elliptical orbiting space vehicles are more erratic and less stable than those of vehicles in circular orbits. For this reason, near-circular orbits are used for most all rendezvous missions, and needless to say, circular orbits greatly simplify the onboard GNC functions for proximity operations. However, the following relative motion analysis does find application when flying rendezvous or proximity operations with space vehicles in elliptical transfer orbits.

As in the circular orbit relative motion analysis, deviations from actual trajectories will begin to accumulate as the range from the LVLH coordinate center becomes large. These differences are due to the approximation for the relative gravitational accelerations, such as those made in Eq. (2.98). Also note that these analyses have been made using the spherical gravitational fields and do not contain the *real world* perturbations due to the gravitational field harmonics.

When the reference space vehicle is in an elliptical orbit, the coefficients (μr^{-3}), ω_o, and $\dot{\omega}_o$, in the differential equations of Eq. (2.150), become variables and are functions of time, and hence the true anomaly as discussed in Sec. 2.2.3. To solve for the relative motion, the following orbital parameters become a necessary part of the propagation methods used:

1) t_o, the LVLH reference orbit epoch time;
2) $M(t_o)$, the orbit mean anomaly at t_o;
3) a, the semimajor axis of the reference orbit;
4) e, the eccentricity of the reference orbit; and
5) $\bar{x}(t_o)$, the initial relative state vector at time t_o, as given in Eq. (2.154).

Note since the solution is to be accomplished by numerical methods, the acceleration components $\bar{f}_i/m$ no longer need to be constants but can also be functions of time.

The coefficients for Eq. (2.150) must be evaluated at each time step as the integration is being performed. The necessary functions of time are repeated here, as presented in Sec. 2.2.3, and are as follows:

$$M(t) = M(t_o) + n(t - t_o) \tag{2.156}$$

where

$$n = \sqrt{\mu}/(a^{3/2}) \tag{2.157}$$

Starting with the mean anomaly $M(t)$ in Eq. (2.81), we have the eccentric anomaly $E(t)$ and then we have the true anomaly $\nu(t)$ directly from Eq. (2.75). The orbit radius is given by

$$r = \frac{a(1 - e^2)}{(1 + e\cos(\nu(t)))} \tag{2.158}$$

The coordinate axis turning rate, which is simply the negative of the rate of change of the true anomaly (orbital rate), is

$$\omega_o = -\sqrt{\frac{\mu}{a(1 - e^2)}}\left[\frac{(1 + e\cos\nu)}{r}\right] \tag{2.159}$$

and its rate of change becomes

$$\dot{\omega}_o = -2(\dot{r}/r)\omega_o \tag{2.160}$$

The rate of change of the orbit's radius magnitude is given by

$$\dot{r} = (e\sin(\nu(t))\sqrt{\mu/(a(1 - e^2))} \tag{2.161}$$

Using these relationships and Eq. (2.150) as the integrand, the ending numerical solution is the LVLH relative motion state $\bar{x}(t)$ at time t.

It is interesting to annotate the preceding discussion of this section by introducing the *local-inertial-velocity* (LIV) coordinates in place of the standard LVLH coordinate frame. The LIV coordinates are defined in a similar manner as the LVLH coordinates of Sec. 1.1.4, but the instantaneous inertial velocity direction is used to define the $\bar{x}_1$ axis as follows:

$$\boldsymbol{i}' = \frac{\boldsymbol{V}}{|\boldsymbol{V}|} \tag{2.162}$$

The $\bar{x}_2$ axis remains perpendicular to the orbital plane, the same as in the LVLH frame definition:

$$\boldsymbol{j}' = -\frac{\boldsymbol{R} \times \boldsymbol{V}}{|\boldsymbol{R} \times \boldsymbol{V}|} \tag{2.163}$$

The $\bar{x}_3$ axis is no longer vertically downward, but becomes

$$\boldsymbol{k}' = \frac{\boldsymbol{i}' \times \boldsymbol{j}'}{|\boldsymbol{i}' \times \boldsymbol{j}'|} \tag{2.164}$$

We have the LIV-to-inertial-(ECI)-coordinate transformation matrix, as in Eq. (1.36):

$$(a) = (\boldsymbol{i}', \boldsymbol{j}', \boldsymbol{k}') = (a^1_\alpha, a^2_\alpha, a^3_\alpha) \tag{2.165}$$

The LIV coordinate frame provides a somewhat better system to visualize relative motion trajectories from space vehicles that are in elliptical orbits. However, the preceding comments concerning the unstable nature of these relative motion trajectories are the same when working with either the LVLH or LIV reference coordinate frames.

When we change to the LIV frame, Eq. (2.148) becomes

$$\bar{R}^i = a^i_\alpha X^\alpha = \begin{pmatrix} \bar{R}_1 \\ \bar{R}_2 \\ \bar{R}_3 \end{pmatrix} = \begin{pmatrix} R \sin\gamma \\ 0 \\ -R\cos\gamma \end{pmatrix} \tag{2.166}$$

where γ is the vehicle flight path angle given in Eq. (2.54). The approximation for the $\bar{P}$ matrix from Eq. (2.98) is changed to

$$(\bar{P}) = -\mu r^{-3} \begin{pmatrix} (1 - 3\sin^2\gamma) & 0 & (3\sin\gamma\cos\gamma) \\ 0 & 1 & 0 \\ (3\sin\gamma\cos\gamma) & 0 & (1 - 3\cos^2\gamma) \end{pmatrix} \tag{2.167}$$

Finally, the LIV relative acceleration components from Eq. (2.147) become

$$
\begin{aligned}
\dot{\bar{v}}_1 &= [\omega_o^2 - \mu r^{-3}(1 - 3\sin^2\gamma)]\bar{x}_1 - (\dot{\omega}_o + 3\mu r^{-3}\sin\gamma\cos\gamma)\bar{x}_3 + 2\omega_o\bar{v}_3 + \frac{\bar{f}_1}{m} \\
\dot{\bar{v}}_2 &= -\mu r^{-3}\bar{x}_2 + \frac{\bar{f}_2}{m} \\
\dot{\bar{v}}_3 &= (\dot{\omega}_o - 3\mu r^{-3}\sin\gamma\cos\gamma)\bar{x}_1 + [\omega_o^2 - \mu r^{-3}(1 - 3\cos^2\gamma)]\bar{x}_3 - 2\omega_o\bar{v}_1 + \frac{\bar{f}_3}{m}
\end{aligned}
\tag{2.168}
$$

where ω_o and $\dot{\omega}_o$ are given in Eqs. (2.159) and (2.160), respectively.

The integrals of Eqs. (2.168) yield the same relative motion in the LIV frame as the integrals of Eqs. (2.150) in the LVLH frame. However, the V-bar crossings, i.e., when the co-orbiting vehicle crosses the velocity direction of the space vehicle, which are shown in the LIV relative motion trajectories, make it possible for the aerospace engineer to better analyze the characteristics of the motion of the co-orbiting vehicle.

2.4.5 Astrodynamics Applications: C-W Guidance Targeting

During rendezvous and docking operations, relatively low velocity approaches are made to the target vehicle, perhaps a space station, to facilitate accurate control and safety from a collision of the two space vehicles. The final rendezvous maneuver generally takes several orbit revolutions after the rendezvous vehicle has been placed in an initial position, i.e., in a near-coplanar orbit several hundred kilometers trailing the target space station. The guidance start time is selected at about two revolutions behind the target vehicle and when the space vehicle is within radar range to begin the rendezvous maneuver. Also, at this time the space vehicle is within the range of accuracy where the C-W equations find excellent application and the onboard range and range rate sensors are capable of acquiring the target space station. The initial relative position and velocity measurements, i.e., t_o, $\bar{\boldsymbol{x}}_o$, and $\bar{\boldsymbol{v}}_o$, can now be used from the onboard sensors and navigation systems.

The follwing analysis by Neily and Jackson[39] can be used by onboard guidance systems to determine the rocket engine firings necessary to maneuver the space vehicle for a safe approach to the target space station. The rendezvous trajectories can be controlled with a number of burns of the rocket engines. For this example, a *two-burn solution* is presented to bring the space vehicle to the desired target relative position and velocity. The changes in relative velocity vectors, i.e., the $\Delta\bar{\boldsymbol{v}}_1$ and $\Delta\bar{\boldsymbol{v}}_2$, are estimated using the following definitions and assumptions:

1) The rendezvous is started with the initial state $(t_o, \bar{\boldsymbol{x}}_o, \bar{\boldsymbol{v}}_o)$.

2) The target state is $(t_t, \bar{\boldsymbol{x}}_t, \bar{\boldsymbol{v}}_t)$ and is generally several hundred meters in front of (along the V-bar line, the line in the velocity direction of the targeted space vehicle) the space station with a relative velocity back toward the space station, similar to the relative motion trajectory as shown in Fig. 2.7. From this vantage point, it is

simpler for braking the relative velocity of the rendezvous space vehicle and at the same time circularizing its orbit at the space station's altitude. The final approach of the rendezvous space vehicle can now be made with good visible contact and ample time to control the vehicle to the docking point. It is also interesting to point out that the target state is also selected so that if the final braking burn is not accomplished, the space vehicle will pass safely below the space station and move ahead and away for another rendezvous attempt, again as depicted in Fig. 2.7.

3) The execution times for the two burns are t_1 and t_2. These two times are generally planned about one orbital period apart, but can depend on the initial separation distance between the two space vehicles at the start of the rendezvous.

With these definitions in mind, the initial state vector is propagated forward to t_1 with the transition matrix formed with $\Delta t = t_1 - t_o$ and the ω_o of the space station orbit using Eq. (B.1) or Eq. (2.153) with the relative accelerations set to zero:

$$\begin{aligned} \bar{\boldsymbol{x}}_1 &= (T_{11})\bar{\boldsymbol{x}}_o + (T_{12})\bar{\boldsymbol{v}}_o \\ \bar{\boldsymbol{v}}_1 &= (T_{21})\bar{\boldsymbol{x}}_o + (T_{22})\bar{\boldsymbol{v}}_o \end{aligned} \tag{2.169}$$

The target state is propagated backward with a second transition matrix formed with $\Delta t = t_2 - t_t$ and the ω_o as in the preceding, and we have the necessary state conditions at t_2:

$$\begin{aligned} \bar{\boldsymbol{x}}_2 &= (T_{11})\bar{\boldsymbol{x}}_t + (T_{12})\bar{\boldsymbol{v}}_t \\ \bar{\boldsymbol{v}}_2 &= (T_{21})\bar{\boldsymbol{x}}_t + (T_{22})\bar{\boldsymbol{v}}_t \end{aligned} \tag{2.170}$$

A third transition matrix is formed to propagate states from t_1 to t_2 with $\Delta t = t_2 - t_1$ and ω_o. We now make use of the partitioned transition matrices of Eqs. (B.2), given in Appendix B. The desired position state computed in Eq. (2.170) at time t_2 can be achieved with a $\Delta\bar{\boldsymbol{v}}_1$, via a rocket engine burn at t_1. This desired position and the natural position states are given here, both at t_2, respectively,

$$\begin{aligned} \bar{\boldsymbol{x}}_2 &= (T_{11})\bar{\boldsymbol{x}}_1 + (T_{12})(\bar{\boldsymbol{v}}_1 + \Delta\bar{\boldsymbol{v}}_1) \\ \hat{\bar{\boldsymbol{x}}}_2 &= (T_{11})\bar{\boldsymbol{x}}_1 + (T_{12})\bar{\boldsymbol{v}}_1 \end{aligned} \tag{2.171}$$

where the necessary position and velocity states at t_1 are computed from Eqs. (2.169). Subtracting the two equations of Eqs. (2.171) and solving for the $\Delta\bar{\boldsymbol{v}}_1$, which will force the trajectory of the space vehicle to the required position state of $\bar{\boldsymbol{x}}_2$ at t_2,

$$\Delta\bar{\boldsymbol{v}}_1 = (T_{12})^{-1}(\bar{\boldsymbol{x}}_2 - \hat{\bar{\boldsymbol{x}}}_2) \tag{2.172}$$

where $\bar{\boldsymbol{x}}_2$ is given in the first equation of Eq. (2.170). The desired velocity state at t_2 to reach the target conditions is given in the second equation of Eq. (2.170), but the new propagated velocity state from t_1 to t_2 becomes

$$\hat{\bar{\boldsymbol{v}}}_2 = (T_{21})\bar{\boldsymbol{x}}_1 + (T_{22})(\bar{\boldsymbol{v}}_1 + \Delta\bar{\boldsymbol{v}}_1) \tag{2.173}$$

The difference between these velocity states is the required $\triangle\bar{v}_2$ at t_2 to send the space vehicle on the trajectory to the target:

$$\triangle\bar{v}_2 = \bar{v}_2 - \hat{\bar{v}}_2 \tag{2.174}$$

The $\triangle\bar{v}_1$ and $\triangle\bar{v}_2$ burn vectors define both the magnitude and the direction of the required rocket engine burns. Although these burns, as computed here, are assumed to occur instantaneously, the actual rocket engine thrust will require a finite time to achieve the velocity changes as required. Hence, the actual ignition times for the burn maneuvers will be scheduled earlier than the burn times computed here, so that the *finite burns* will achieve the same effects on the trajectory of the space vehicle.

Using the ideal rocket equations from Sec. 2.1.4, the weight of the propellants required for the rendezvous can be estimated. Equation (2.34) computes the change in space vehicle weight or the weight of the propellant required for the burn, when the weight of the space vehicle, the I_{sp} of the rocket engine system, and the $\triangle\bar{v}$ are known.

Conclusion

In chapter two I have presented the reader with the analysis and supporting mathematical relationships dealing with the motion of the point mass. This is again, in the theme of the text, directed to form the knowledge base for the reader that is necessary for the solutions of actual aerospace problems. The fundamental concepts of gravitation and the motion of a point mass in two-body space have been applied to the Keplerian solution for the *Two-body problem of astrodynamics*. I would like to reemphasize here that the Keplerian solution is only an approximation for the real world motion of a space vehicle. However, it is very important to the reader, because it defines all of the basic orbital parameters that are referred to in the analysis of real orbits.

I have introduced the reader to a four-degree-of-freedom (4DOF) airplane motion simulation which is performed in Earth-based or station coordinates. In this analysis I have defined the airplane dynamic parameters and the fist order linear force coefficient equations to estimate the aerodynamic forces on the airframe.

Point mass motion equations have been derived in space-based coordinates, LVLH coordinates, body-axis coordinates, and now in Earth-based coordinates. I would like to emphasize that both the Earth-based and the LVLH coordinate frames are not what I call *inertial coordinate frames*. In conclusion, I define an inertial coordinate frame as one where we attach Cartesian coordinate axes to a mass which is in motion in gravitational space, but; 1) there are *no contact forces* accelerating that mass and; 2) the attached coordinate axes are *stationary with respect to the distant stars*.

I have derived the basic equations for onboard space vehicle mircogravity environment. This gives the reader a method for estimating the accelerations felt by a mass or an accelerometer located anywhere onboard the space vehicle.

Space vehicle relative motion equations for a rendezvous vehicle are derived in the LVLH coordinate frame. When the space vehicle is in a circular orbit the relative motion equations have a closed form approximating solution. In aerospace terminology this solution is called simply the C-W solution. Although the C-W solution is an approximation for the real motion of the rendezvous vehicle, the aerospace engineer has found the relative motion predictions to be very accurate out to several kilometers. The C-W equations are accurate enough to be used in the onboard rendezvous guidance software.

In contrast with the space vehicles in circular orbits, the relative motion equations for space vehicles in elliptical orbits are shown not to have a closed form solution. For example, this type of relative motion would occur when a rendezvous is necessary with a space vehicle in an elliptical transfer orbit. In this case the onboard software system must numerically integrate the relative motion equations to solve for the motion of the rendezvous vehicle.

3
N-Body Gravitational Space and Rigid Body Motion

Introduction

In this chapter I discuss the mathematical descriptions of a system of N mass points. I have grouped these analyses into two general areas for study: 1) the motion of a system of mass points which form a rigid body and define the equations for rigid body dynamics, and 2) the motion of N independent mass points, as in planetary motion, which becomes the N-Body problem of astrodynamics.

I begin by defining the center of mass of a rigid body and the constraints that are placed on the individual mass particles making up the rigid body. The kinetic energy of the rigid body is derived in tensor form, and in doing so the moment of inertia matrix is shown to be a second order tensor. Using the summation convention notation I derive the equations of motion for a rigid body in the conservative gravity field of a large planetary mass. With this analysis I derive Euler's rotational equations of motion, where I have included the effects of balanced spinning subbodies, such as the rapidly spinning parts of a jet engine.

The slight displacement of the center-of-gravity from the center-of-mass of a rigid body due to a gravity gradient is developed in matrix form. This displacement causes the space vehicle to feel the rotational effects due to the gravity gradient torques. The rigid body gravity gradient torques are also developed and are shown to be a function of the body-axis moment of inertia tensor.

Computing the orbits of a space vehicle in cislunar space is the theme of an aerospace application of this chapter. The method outlined and described uses the sun, Earth, and moon locations as given in *The Astronomical Almanac*. These locations are given in the Earth-centered-inertial coordinate frame and are given to sufficient accuracy for use in this engineering motion simulation. This method simplifies the problem of precisely locating the Earth-sun-moon point masses so that the engineer can concentrate on the motion of families of space vehicle orbits to define the desired cislunar trajectories.

3.1 *N*-Body Mass Systems: Mathematical Descriptions

3.1.1 Center of Mass

The center of mass of a system of mass particles or of a body is a single point, where the total mass of that body could be considered to be located. We can sum on all mass particles that make up the system of particles or the body and define the point X_i, such that

$$m_\rho x_i^\rho = MX_i \tag{3.1}$$

where the notation x_i^ρ, as in Sec. 2.1.1, is not a second-order tensor, but represents the x_i vector components of the first-order tensor, i.e., the inertial position vector to the ρth mass particle, and where

$$M = \sum_\rho^n m_\rho \tag{3.2}$$

is the total mass of the body. We can locate the ρth particle of mass by Eq. (1.14) and sum on all particles of mass, which will define the body:

$$m_\rho x_i^\rho - m_\rho X_i^\rho = m_\rho a_i^\alpha \bar{x}_\alpha^\rho \tag{3.3}$$

If the center of mass point can be found, then the X_i can be factored from the second term of Eq. (3.3), leaving the sum of the mass particles, and based on Eq. (3.1),

$$m_\rho a_i^\alpha \bar{x}_\alpha^\rho = 0 \tag{3.4}$$

Therefore,

$$X_i = \frac{m_\rho x_i^\rho}{M} \tag{3.5}$$

defines the center of mass and also requires the condition defined by Eq. (3.4). Differentiating Eq. (3.4) and using Eq. (1.222), we have additional conditions required from the center of mass definition:

$$m_\rho a_i^\beta \bar{v}_\beta^\rho = m_\rho a_i^\beta \left(\bar{W}_\beta^\alpha \bar{x}_\alpha^\rho + \bar{v}_\beta^\rho\right) = 0 \tag{3.6}$$

Using this constraint and by differentiating Eq. (3.3), we have a similar equation to Eq. (3.5) for the velocity of the center of mass:

$$V_i = \frac{m_\rho v_i^\rho}{M} \tag{3.7}$$

In this text the following rigid body definitions apply: 1) *a body where all of its mass particles have body relative velocities of zero*, i.e., all $\bar{v}_\beta^\rho = 0$, in Eq. (3.6); or 2) *a body where some of its mass particles have non-zero body relative velocities as a result of balanced spinning subbodies* (such as the rapidly spinning parts of a jet engine). In both of these cases, the center of mass definition in Eq. (3.5) and (3.6) are still true. Hence, *the body relative center of mass is not changed as a result of the presence of rotating subbodies*. When rotating subbodies are included in the rigid body, the rotational dynamics of the total body will be shown to be affected.

We can define the location of the mass particles of only the spinning parts in body axis coordinates as follows:

$$\bar{x}_i^\sigma = b_i^\alpha \bar{\bar{x}}_\alpha^\sigma + \bar{X}_i \tag{3.8}$$

Accounting for the spinning subbody, Eq. (3.4) becomes

$$m_\rho a_i^\alpha \bar{x}_\alpha^\rho + m_\sigma a_i^\alpha \left(b_\alpha^\gamma \bar{\bar{x}}_\gamma^\sigma + \bar{X}_\alpha^\sigma\right) = 0 \tag{3.9}$$

where the indicated sums can be divided into two parts: 1) on ρ for the body axis stationary mass particles and 2) on σ for the mass particles of the spinning subbody. $\bar{X}_\alpha^\sigma$ is the location of the center of mass of the spinning subbody in body relative coordinates. Each of the σ spinning mass particles will have a body axis relative velocity of

$$\bar{v}_i^\sigma = b_i^\beta \bar{\bar{W}}_\beta^\gamma \bar{\bar{x}}_\gamma^\sigma \tag{3.10}$$

where $\bar{\bar{W}}_\beta^\gamma$ is the body axis rotation rate matrix of the spinning subbody.

If the system of mass particles is not stationary in body coordinates and their motion is not due to a defined spinning subbody, we have what is called a *flexible body*, and the center of mass definitions do not apply. The body relative position of the center of mass of a flexible body moves when the body's structure deforms. The motion of flexible bodies is beyond the scope of this text.

Differentiating Eq. (3.6) again, we have yet another constraint from the center of mass definition:

$$m_\rho a_i^\gamma \bar{a}_\gamma^\rho = 0 \tag{3.11}$$

where the $\bar{a}_\gamma^\rho$ are the acceleration components as discussed in Eq. (1.232).

3.1.2 N-Body Gravitational Space

We now consider a system of N point masses in the vicinity of an inertial coordinate frame. In Sec. 2.2, the gravity potential was considered to be from a single gravitating mass located at the center of the inertial coordinate frame. This gravity field is spherical in nature and forms the gravitational space environment for the analysis of the two-body problem of astrodynamics. On the other hand, in N-body gravitational space, the gravity potential at any point is due to the gravity potential from n gravitating masses that form the gravitational environment for the motion analysis. N-body gravitational space is not spherical, and hence motion analysis is considerably more complicated. The total energy of the system is the sum of the energy of the single mass particles, as in Eq. (2.35), and can now be written

$$E = (1/2)m^\rho v_\rho^2 - Gm^\rho m_\gamma (|\boldsymbol{r}^\gamma - \boldsymbol{r}_\rho|)^{-1} \tag{3.12}$$

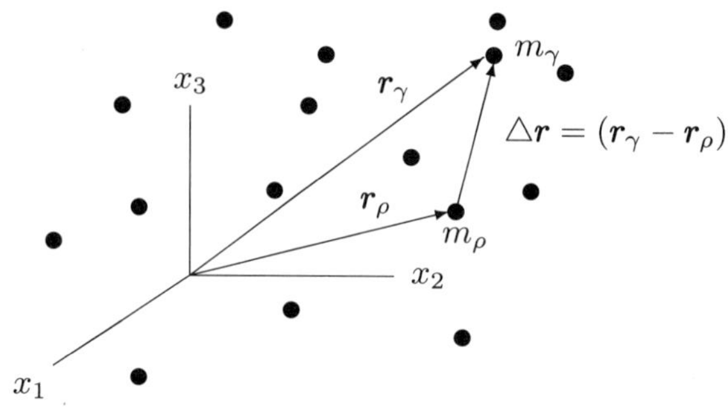

Fig. 3.1 *N*-body gravitational space.

where $\rho = 1, \ldots, n$ and $\gamma \neq \rho$, v_ρ is the inertial velocity magnitude of the ρth mass, and the $\boldsymbol{r}_\rho$ and $\boldsymbol{r}^\gamma$ are inertial vectors to their respective mass particles, as shown in Fig. 3.1.

Unlike in Sec. 2.2.1, each of the ρth mass particles could be free to move or constrained within a rigid body, but nevertheless each will influence the gravitational field of the other, as stated in Newton's law of gravitation. To determine the differential equations of motion for each mass in the system, we can use a similar procedure as in Sec. 2.2.2, but we must realize that the derivative of the potential now results in pairs of terms producing the accelerations along each line connecting the gravitating mass pairs. Differentiating Eq. (3.12) with respect to time, we have

$$\begin{aligned}\frac{\mathrm{d}E}{\mathrm{d}t} &= \boldsymbol{f}^\rho \cdot \boldsymbol{v}_\rho \\ &= m^\rho(\dot{\boldsymbol{v}}_\rho \cdot \boldsymbol{v}_\rho) + Gm^\rho m_\gamma(|\boldsymbol{r}^\gamma - \boldsymbol{r}_\rho|)^{-3}[(\boldsymbol{r}^\gamma - \boldsymbol{r}_\rho)\cdot \boldsymbol{v}^\gamma - (\boldsymbol{r}^\gamma - \boldsymbol{r}_\rho)\cdot \boldsymbol{v}_\rho]\end{aligned} \tag{3.13}$$

where the rate of change of energy in the system is equal to the sum of all contact forces acting on each of the mass particles. By equating the terms of the same vector dot products, we have the equations of motion for each mass particle, which is free to move with unconstrained motion. For the ith mass particle, we have

$$\boldsymbol{f}_i = m_i\dot{\boldsymbol{v}}_i - Gm_i m^\gamma(|\boldsymbol{r}_\gamma - \boldsymbol{r}_i|)^{-3}(\boldsymbol{r}_\gamma - \boldsymbol{r}_i) \tag{3.14}$$

where $\gamma = 1, \ldots, n$ but $\gamma \neq i$.

These equations, from the foundations of classical mechanics, form the defining relationships for many aerospace and astrodynamic applications. The complete generality of mass distributions and velocities implied in Eqs. (3.12–3.14) make it necessary to define constraints and specify the nature of the gravitating mass points for each analysis application. For instance, four different physical systems of mass particles can be specified and approximated using these equations as the starting assumptions:

1) The motion of systems of mass particles that form a rigid body in gravitational space; rigid body dynamics is discussed in Secs. 3.1.1–3.1.6 and in all sections of Sec. 3.2.

2) The application to the motion of n-point masses used in the analysis of planetary and/or space vehicle orbital paths; n-body motion is applied to Earth-lunar orbits in Sec. 3.1.7.

3) The analysis of and approximations for the nonspherical gravitational fields of planetary masses; approximating gravitational environments is discussed in Secs. 4.1.1–4.1.3.

4) Analysis of the gravitational interactions between astronomical bodies causing perturbations of both their translational and rotational motion; the analysis of the interacting gravitational torques and orbital motion of large rigid bodies or flexible bodies; the analysis of the motion of these systems of mass particles is not covered in this text.

3.1.3 Kinetic Energy of a Rigid Body and the Moment-of-Inertia Tensor

Consider now a rigid body consisting of n mass points like those shown in Fig. 3.1. Using only the kinetic energy term of Eq. (3.12), we can use the vector dot product, $\boldsymbol{v}_\rho \cdot \boldsymbol{v}_\rho$, for the velocity magnitude squared v_ρ^2. However, a summation sign is necessary in the notation to indicate this dot product. The components of inertial velocity vector of each mass point in the system of particles is given in Eq. (1.223), and the a_i^γ are the components of the body-axis-to-inertial transformation matrix. The kinetic energy becomes

$$E_k = (1/2)m_\rho \sum_{i=1}^{3} \left[(a_i^\gamma \bar{v}_\gamma^\rho + V_i^\rho)(a_i^\sigma \bar{v}_\sigma^\rho + V_i^\rho)\right] \tag{3.15}$$

Using the center of mass conditions in Eq. (3.6) and the fact that the center of mass velocity will factor from the summation of those mass particles, leaving the total rigid body mass M, we have

$$E_k = \sum_{i=1}^{3} \left[a_i^\gamma a_i^\sigma ((1/2)m_\rho \bar{v}_\gamma^\rho \bar{v}_\sigma^\rho) + (1/2)MV_iV_i\right] \tag{3.16}$$

The dot product in the first term is invariant during coordinate transformations, as in Eq. (1.64), and the kinetic energy reduces further to

$$E_k = \sum_{i=1}^{3} ((1/2)m_\rho \bar{v}_i^\rho \bar{v}_i^\rho) + (1/2)MV^2 \tag{3.17}$$

where the first term is the kinetic energy due to rotation, and the last term is simply the kinetic energy due to the translational motion of the total mass. Continuing with the expansion of the first term where $\bar{v}_i^\rho = \bar{W}_i^\beta \bar{x}_\beta^\rho + \bar{v}_i^\rho$, from Eq. (1.222), we can require that the $\bar{v}_i^\rho$ are all zero for the rigid body case. We can also rearrange the first terms of Eq. (1.222) as follows:

$$\bar{W}_i^{\beta}\bar{x}_\beta^\rho = \begin{pmatrix} 0 & -\bar{\omega}_3 & \bar{\omega}_2 \\ \bar{\omega}_3 & 0 & -\bar{\omega}_1 \\ -\bar{\omega}_2 & \bar{\omega}_1 & 0 \end{pmatrix} \begin{pmatrix} \bar{x}_1^\rho \\ \bar{x}_2^\rho \\ \bar{x}_3^\rho \end{pmatrix} = \begin{pmatrix} 0 & \bar{x}_3^\rho & -\bar{x}_2^\rho \\ -\bar{x}_3^\rho & 0 & \bar{x}_1^\rho \\ \bar{x}_2^\rho & -\bar{x}_1^\rho & 0 \end{pmatrix} \begin{pmatrix} \bar{\omega}_1 \\ \bar{\omega}_2 \\ \bar{\omega}_3 \end{pmatrix} \tag{3.18}$$

Borrowing the summation notation form, we have

$$\bar{W}_i^{\beta}\bar{x}_\beta^\rho = \bar{\mathcal{X}}_i^{\rho\gamma}\bar{\omega}_\gamma \tag{3.19}$$

The notation $\bar{\mathcal{X}}_i^{\rho\gamma}$ is not a three-dimensional matrix element, but it represents the ρth mass particle in the (i, γ)th element location of the $(\bar{\mathcal{X}}^\rho)$ matrix. Using this expression in Eq. (3.13),

$$E_k = \sum_{i=1}^{3}\left((1/2)m_\rho \bar{\mathcal{X}}_i^{\rho\beta}\bar{\mathcal{X}}_i^{\rho\gamma}\bar{\omega}_\beta\bar{\omega}_\gamma\right) + (1/2)MV^2 \tag{3.20}$$

During further examination of the first term, a very important group of sums occur that appear repeatedly in the dynamics of rigid bodies:

$$\bar{I}^{\beta\gamma} = \sum_{i=1}^{3} m_\rho \bar{\mathcal{X}}_i^{\rho\beta}\bar{\mathcal{X}}_i^{\rho\gamma} \tag{3.21}$$

and written out in matrix form,

$$(\bar{I}) = \begin{pmatrix} m_\rho(\bar{x}_2^\rho\bar{x}_2^\rho + \bar{x}_3^\rho\bar{x}_3^\rho) & -m_\rho\bar{x}_1^\rho\bar{x}_2^\rho & -m_\rho\bar{x}_1^\rho\bar{x}_3^\rho \\ -m_\rho\bar{x}_2^\rho\bar{x}_1^\rho & m_\rho(\bar{x}_1^\rho\bar{x}_1^\rho + \bar{x}_3^\rho\bar{x}_3^\rho) & -m_\rho\bar{x}_2^\rho\bar{x}_3^\rho \\ -m_\rho\bar{x}_3^\rho\bar{x}_1^\rho & -m_\rho\bar{x}_3^\rho\bar{x}_2^\rho & m_\rho(\bar{x}_1^\rho\bar{x}_1^\rho + \bar{x}_2^\rho\bar{x}_2^\rho) \end{pmatrix} \tag{3.22}$$

This matrix is called the body-axis-moment-of-inertia tensor and by definition, as shown in this analysis, is referenced to the center of mass of the rigid body. The tensor character of the moment-of-inertia matrix will be demonstrated by how it transforms into other coordinate systems and is discussed in the next section of the text. The moment-of-inertia tensor effectively describes the mass distribution of a rigid body and is used throughout the analysis of rotational dynamics. Using the moment-of-inertia tensor, the kinetic energy of a rigid body becomes

$$E_k = (1/2)\bar{\omega}^\alpha \bar{I}_\alpha^\beta \bar{\omega}_\beta + (1/2)MV^2 \tag{3.23}$$

The kinetic energy of the rigid body is the sum of the *rotational kinetic energy* and the *kinetic energy due to the translation* of its center of mass.

3.1.4 Transformations of the Moment-of-Inertia Tensor

If we have a rigid body in inertial space, the moment-of-inertia tensor can be defined in reference to any coordinate center and any body axis orientation in space. However, definition from the center of mass, as in Eq. (3.22), finds most application and meaning in rigid body dynamics. For the analysis of this section, we define the moment-of-inertia tensor in reference to the coordinate center of the inertial frame, and in a similar manner to Eq. (3.22), we have

$$(I) = \begin{pmatrix} m_\rho(x_2^\rho x_2^\rho + x_3^\rho x_3^\rho) & -m_\rho x_1^\rho x_2^\rho & -m_\rho x_1^\rho x_3^\rho \\ -m_\rho x_2^\rho x_1^\rho & m_\rho(x_1^\rho x_1^\rho + x_3^\rho x_3^\rho) & -m_\rho x_2^\rho x_3^\rho \\ -m_\rho x_3^\rho x_1^\rho & -m_\rho x_3^\rho x_2^\rho & m_\rho(x_1^\rho x_1^\rho + x_2^\rho x_2^\rho) \end{pmatrix} \tag{3.24}$$

The system of mass particles of the rigid body are each located in the inertial coordinate frame by

$$x_i^\rho = a_i^\beta \bar{x}_\beta^\rho + X_i \tag{3.25}$$

where the a_i^β are the components of the body-axis-to-inertial transformation matrix, and the X_i are the vector components to the center of mass of the rigid body. By substituting Eq. (3.25) into Eq. (3.24) for each element of the moment-of-inertia tensor, utilizing the transformation matrix identities of Eqs. (1.40) and (1.41) and using the center of mass constraints in Eq. (3.4), we have

$$I_{\alpha\beta} = a_{\alpha\gamma} a_{\beta\sigma} \bar{I}^{\gamma\sigma} + M \begin{pmatrix} X_2^2 + X_3^2 & -X_1X_2 & -X_1X_3 \\ -X_2X_1 & X_1^2 + X_3^2 & -X_2X_3 \\ -X_3X_1 & -X_3X_2 & X_1^2 + X_2^2 \end{pmatrix} \tag{3.26}$$

Here we have carried out the summations on ρ for all of the mass particles of the rigid body, and the $\bar{I}^{\gamma\sigma}$ are the components of the moment-of-inertia tensor relative to the center of mass of the rigid body itself, as given in Eq. (3.22). The first expression on the right-hand side of Eq. (3.26) is the transformation of the *moment-of-inertia tensor due to the rotation* of the rigid body from the reference coordinate frame and is equivalent to $I = (a)(\bar{I})(a)^T$ in matrix notation. This term can also be written using the equalities given in Eqs. (1.16) and (1.17), as follows:

$$I_i^j = \bar{I}_\alpha^\beta a_i^\alpha a_\beta^j = \bar{I}_\alpha^\beta \frac{\partial x_i}{\partial \bar{x}_\alpha} \frac{\partial \bar{x}^j}{\partial x_\beta} \tag{3.27}$$

Comparing this transformation equation to the general form of a tensor, as shown by Eq. (1.23), we see that the moment-of-inertia matrix is indeed a second-order tensor. The second term of Eq. (3.26) is due to the *translation of the total mass* to a location other than the orginal inertial coordinate center.

The *Parallel Axis Theorem* (Ref. 21, pp. 225, 425) states that the moment-of-inertia tensor about any given axis is equal to the moment of inertia about a parallel axis through the center of mass, plus the moment of inertia about the given axis as if all the mass of the body were located at the center of mass. The Parallel Axis

Theorem is a special case of the more general case described in Eq. (3.26). For the parallel axis case, the transformation matrix a becomes the unit matrix, i.e., there has been no rotation of the rigid body, and its axes remain parallel to or the same as the original angular orientation.

Equation (3.26) makes it possible for the aerospace engineer[40] to estimate the moment-of-inertia tensor of the total aircraft or spacecraft body by summing the inertia tensors of each subbody based on its angular orientation and its center of mass location with respect to a common coordinate center of the total aerospace vehicle.

3.1.5 Total Angular Momentum of a Rigid Body

The angular momentum of a mass particle in inertial coordinates is given in Eq. (2.22). The inertial coordinates for the position and velocity for each particle of mass are given in Eqs. (1.168) and (1.221), respectively. In this analysis we assume that the center of mass of the rigid body is at X_i, with an inertial velocity of V_i. The body-axis-to-inertial transformation matrix is given by the a_i^α components. In most rigid body analysis applications, the mass particles of the rigid body are stationary in body axis coordinates, i.e., $\bar{v}_\beta^\rho = 0$. However, in the more general case where rotating subbodies need to be included, we can write Eq. (1.221) using the right-hand side of Eq. (3.19) as follows:

$$v_i^\rho = a_i^\beta (\bar{\mathcal{X}}_\beta^{\rho\gamma} \bar{\omega}_\gamma + \bar{v}_\beta^\rho) + V_i^\rho \tag{3.28}$$

The non-zero $\bar{v}_\beta^\rho$ body axis velocity components will appear in the summations of the mass particles of the rotating subbodies. Using this expression for velocity and Eq. (1.168) for position, substituting into Eq. (2.22) and applying the center of mass condition of Eq. (3.6), we have

$$\begin{aligned}
L_1 &= a_2^\alpha a_3^\beta \left[m_\rho (\bar{x}_\alpha^\rho \bar{\mathcal{X}}_\beta^{\rho\gamma} - \bar{x}_\beta^\rho \bar{\mathcal{X}}_\alpha^{\rho\gamma}) \bar{\omega}_\gamma + (\bar{x}_\alpha^\rho \bar{v}_\beta^\rho - \bar{x}_\beta^\rho \bar{v}_\alpha^\rho) \right] + M(X_2 V_3 - X_3 V_2) \\
L_2 &= a_3^\alpha a_1^\beta \left[m_\rho (\bar{x}_\alpha^\rho \bar{\mathcal{X}}_\beta^{\rho\gamma} - \bar{x}_\beta^\rho \bar{\mathcal{X}}_\alpha^{\rho\gamma}) \bar{\omega}_\gamma + (\bar{x}_\alpha^\rho \bar{v}_\beta^\rho - \bar{x}_\beta^\rho \bar{v}_\alpha^\rho) \right] + M(X_3 V_1 - X_1 V_3) \\
L_3 &= a_1^\alpha a_2^\beta \left[m_\rho (\bar{x}_\alpha^\rho \bar{\mathcal{X}}_\beta^{\rho\gamma} - \bar{x}_\beta^\rho \bar{\mathcal{X}}_\alpha^{\rho\gamma}) \bar{\omega}_\gamma + (\bar{x}_\alpha^\rho \bar{v}_\beta^\rho - \bar{x}_\beta^\rho \bar{v}_\alpha^\rho) \right] + M(X_1 V_2 - X_2 V_1)
\end{aligned} \tag{3.29}$$

The total mass terms, i.e., the M terms, are the elements of the angular momentum vector in the inertial frame due to motion of the center of mass of the rigid body. Expanding the sums on α and β in the first term, notice that the $\bar{x}_\alpha^\rho \bar{\mathcal{X}}_\beta^{\rho\gamma} - \bar{x}_\beta^\rho \bar{\mathcal{X}}_\alpha^{\rho\gamma}$ terms will be zero when $\alpha = \beta$. Using Eqs. (1.49), where each element of the transformation matrix is equal to its minor, and all other sums of the first terms, a very interesting relationship is shown when the mass particles are summed:

$$\begin{aligned}
\bar{I}_1^\gamma &= m_\rho (\bar{x}_2^\rho \bar{\mathcal{X}}_3^{\rho\gamma} - \bar{x}_3^\rho \bar{\mathcal{X}}_2^{\rho\gamma}) \\
\bar{I}_2^\gamma &= m_\rho (\bar{x}_3^\rho \bar{\mathcal{X}}_1^{\rho\gamma} - \bar{x}_1^\rho \bar{\mathcal{X}}_3^{\rho\gamma}) \\
\bar{I}_3^\gamma &= m_\rho (\bar{x}_1^\rho \bar{\mathcal{X}}_2^{\rho\gamma} - \bar{x}_2^\rho \bar{\mathcal{X}}_1^{\rho\gamma})
\end{aligned} \tag{3.30}$$

where the $\bar{I}$ matrix is the body-axis-moment-of-inertia tensor of Eq. (3.22). If we momentarily do not include the rotating subbodies in the summations indicated in Eqs. (3.29), the total angular momentum vector of the rigid body in inertial coordinates becomes

$$
\begin{aligned}
L_1 &= a_1^\beta \bar{I}_\beta^\gamma \bar{\omega}_\gamma + M(X_2V_3 - X_3V_2) \\
L_2 &= a_2^\beta \bar{I}_\beta^\gamma \bar{\omega}_\gamma + M(X_3V_1 - X_1V_3) \\
L_3 &= a_3^\beta \bar{I}_\beta^\gamma \bar{\omega}_\gamma + M(X_1V_2 - X_2V_1)
\end{aligned} \tag{3.31}
$$

Using the notation in Eq. (2.24), we have the total angular momentum vector in the inertial frame:

$$L_i = a_i^\alpha \bar{I}_\alpha^\gamma \bar{\omega}_\gamma + M\Upsilon_i^\beta X_\beta \tag{3.32}$$

The angular momentum vector of the rotating body in the body axis coordinates is

$$\bar{L}_\alpha = \bar{I}_\alpha^\gamma \bar{\omega}_\gamma \tag{3.33}$$

and finally Eq. (3.32) simply becomes

$$L_i = a_i^\alpha \bar{L}_\alpha + M\Upsilon_i^\beta X_\beta \tag{3.34}$$

The total angular momentum vector of a rigid body in the inertial frame is the sum of the *angular momentum due to the rotation* of the rigid body and the *angular momentum due to the translational motion* of its center of mass.

To include the momentum of the rotating subbodies, we return to Eqs. (3.29) and expand the $\bar{v}_\beta^\rho$ terms using the same analysis procedures as we did in Eqs. (3.30) and (3.31). The angular momentum of the rotating subbodies appear, and the total angular momentum vector of Eq. (3.34) becomes

$$L_i = a_i^\alpha (\bar{L}_\alpha + b_\alpha^\beta \bar{\bar{L}}_\beta) + M\Upsilon_i^\beta X_\beta \tag{3.35}$$

The angular momentum in the body axis coordinate frame, with the rotating subbody, displays the tensor character of angular momentum, and Eq. (3.33) is rewritten as follows:

$$\bar{L}_\alpha = \bar{I}_\alpha^\gamma \bar{\omega}_\gamma + b_\alpha^\beta \bar{\bar{I}}_\beta^\delta \bar{\bar{\omega}}_\delta \tag{3.36}$$

where $\bar{\bar{I}}_\beta^\delta$ is the moment-of-inertia tensor of the rotating subbody in its coordinate frame, and the $\bar{\bar{\omega}}_\delta$ is the rotation rate vector of the subbody relative to the rigid body axis frame. Note that the moment-of-inertia tensor of a balanced rotating subbody will have diagonal matrix form, i.e., the off-diagonal matrix elements will be equal to zero.

It is interesting to note that the transformation of the moment-of-inertia tensor, as shown by Eq. (3.26) of the previous section, can again be easily verified by using Eq. (3.34) in matrix notation and the following analysis:

$$L = (I)\omega = (a)(\bar{I})\bar{\omega} \tag{3.37}$$

The angular velocity vector in the inertial frame is given by Eq. (1.187), namely, $\omega = (a)\bar{\omega}$, and Eq. (3.37) becomes

$$((I)(a))\bar{\omega} = ((a)(\bar{I}))\bar{\omega} \tag{3.38}$$

and

$$I = (a)(\bar{I})(a)^T \tag{3.39}$$

Again, we recognize the similarity transformation (Ref. 41, p. 36) that transforms the moment-of-inertia tensor in body axis coordinates into the inertial frame. Therefore, we have an equvalent matrix solution for the transformation of the moment-of-inertia tensor as given in tensor notation by the first expression of Eq. (3.26).

3.1.6 Potential Energy of a Rigid Body

For this analysis, we can represent a planet, such as Earth, as a large point mass, with $\mu = GM_p$, and assume that the point mass is also the center of the inertial coordinate frame. Similar to Sec. 2.2.1, the space vehicle is now a system of non-homogeneous mass particles, where the radius distance to all of the mass particles from the center of mass of the space vehicle is *very small* compared to the vehicle's distance from the planet's center. We assume also that the presence of and motion of the space vehicle has absolutely no effect on the motion of the planet itself. The potential energy of the rigid body, from Eq. (3.12), becomes simply

$$E_p = -\mu m^\rho (|\boldsymbol{r}_\rho|)^{-1} \tag{3.40}$$

where components of the $\boldsymbol{r}_\rho$ are given by $x_i^\rho = a_i^\alpha \bar{x}_\alpha^\rho + X_i$. As in Sec. 3.1.3, the a_i^α are the components of the body-axis-to-inertial transformation matrix, and the X_i are the inertial vector components to the center of mass of the rigid body. Using this and Eq. (1.66) in Eq. (3.40), we have

$$E_p = -\mu m^\rho \left(R^2 + \bar{r}_\rho^2 + 2\tilde{\boldsymbol{r}}_\rho \cdot \boldsymbol{R}\right)^{-(1/2)} \tag{3.41}$$

where $\bar{r}_\rho$ is the radius distance from the center of mass of the space vehicle to the ρth mass particle, and the components of $\tilde{\boldsymbol{r}}_\rho$ are $a_i^\alpha \bar{x}_\alpha^\rho$. The location of the center of mass in the inertial frame is $\boldsymbol{R}$. Unfortunately, Eq. (3.41) does not present closed solutions for non-homogeneous bodies. However, we can find approximating solutions. For example, the square-root term can be estimated very accurately using the Legendre polynomials as shown in Ref. 42 (p. 489). If we use only the first three terms of the Legendre expansion and perform the summations on ρ for all mass particles representing the rigid body, we have

$$E_p \approx -\mu M \left(\frac{1}{R}\right) - \left(\frac{\mu}{2R^3}\right)(\bar{T}_R - 3\bar{\bar{I}}_{11}) \tag{3.42}$$

where M is the total mass of the vehicle. Summations for the $m^\rho(\bar{r}_\rho^2)$ term forms one-half of the *trace* of the body-axis-moment-of-inertia matrix, i.e., $\bar{T}_R = \bar{I}_{11} + \bar{I}_{22} + \bar{I}_{33}$. The $\bar{\bar{I}}_{11}$ is the (1,1) element of the vehicle's body-axis-moment-of-inertia matrix $\bar{I}$, which has been transformed to the axes defined by the radius vector $\boldsymbol{R}$ in the body axis frame. This term is formed from the summation $m^\rho(\bar{r}_\rho^2 \sin^2\theta_\rho)$ during the expansion process. The $\bar{\bar{I}}$ matrix is given by $(c)^T(\bar{I})(c)$, as discussed in Sec. 3.1.4. The c matrix defines the transformation $\bar{x} = (c)\bar{\bar{x}}$, where the $\bar{\bar{x}}_1$ axis is defined by $\boldsymbol{R}$ in the body axis coordinates.

Because the space vehicle is very small compared to the planetary mass, we find the magnitude of the second term of Eq. (3.42) almost insignificant. For example, with a vehicle about 50 ft in length and with a mass of 953 slugs located near the Earth's surface, we have:

1) the potential energy from the first term is -0.64105×10^{12} ft · 16,

2) the term $\mu/(2R^3)$ has a magnitude of 0.76926×10^{-06},

3) the trace of the moment-of-inertia matrix in the second term has the magnitude of $0.49314 \times 10^{+06}$, and

4) the term $3\bar{\bar{I}}_{11}$ is nearly the same magnitude as the trace.

These computations are a function of space vehicle attitude, which determines the inertial radius vectors $\boldsymbol{r}_\rho$ to each of the mass particles representing the vehicle. The approximating Eq. (3.42) includes these orientation effects represented by the $\bar{\bar{I}}_{11}$ term. However, again most space vehicles are relatively small, and these effects are not very significant. In conclusion, the potential energy of a space vehicle of mass M is simply

$$E_p \approx -\mu M\left(\frac{1}{R}\right) \tag{3.43}$$

3.1.7 Aerospace Applications: Computing Earth-Lunar Orbits

This application suggests a simple method to simulate the orbital motion of space vehicles in the three-body, sun-Earth-lunar gravitational environment. The engineering analysis and planning for lunar missions require studies of the families of space vehicle orbits in the sun, Earth, and moon gravitational space. From the family of orbits, desired orbits are selected and used in GNC systems to fly the space vehicle on optimum trajectories designed for safety and fuel efficiency.

Using the differential equations of motion given in Eq. (3.14), we can create the multibody gravitational space for the motion of orbiting space vehicles. In this example, however, only three astronomical bodies, namely the sun, Earth, and moon, are used to approximate the orbital dynamics of the space vehicle. The multibody motion problem has no known closed-form solutions, but can be solved by numerical methods. The space vehicle's mass is so small that it has no effect on the orbiting astronomical bodies.

We know precisely the orbits of the Earth and moon, from Ref. 5 or from the Jet Propulsion Laboratory (JPL) Ephemeris Data Files,[25] as they orbit the sun. In this analysis, we simply use the solar and lunar positions as given in Ref. 5 (pp. C24, D22, respectively). These approximate solar and lunar vectors are given in

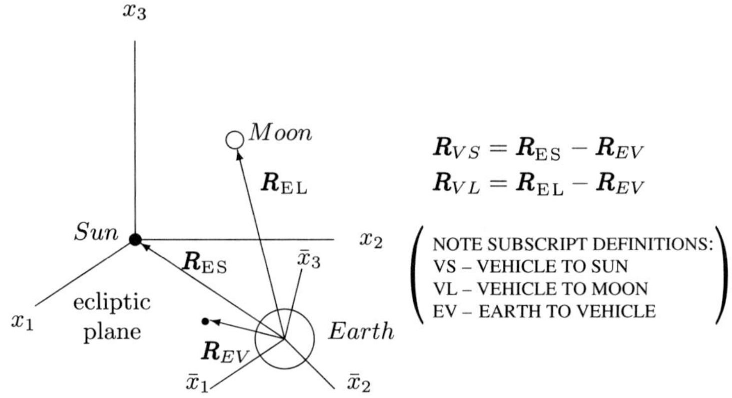

Fig. 3.2 The space vehicle in sun, Earth and moon gravitational space.

the Earth-centered, inertial (ECI) frame, referencing the Earth equatorial plane of date (sometimes refered to as TOD, meaning, true-of-date). Figure 3.2 does not depict an actual Earth and moon position configuration, but it does help visualize vectors relating the sun, Earth, and moon for this analysis of the motion of the space vehicle. In Fig. 3.2 the x_1 and $\bar{x}_1$ axes are defined parallel to each another and are directed precisely to the vernal equinox of date, as discussed in Sec. 1.2.6. The *ecliptic plane* is defined to be the plane of the Earth's orbit around the sun, and the angular momentum vector of the Earth's orbit defines the x_3 axis. The motion of the sun and planets are described using this inertial ecliptic coordinate frame. However, in this analysis application, we will transform the motion into the ECI frame where the differential equations of motion will be written and the motion integrals will be performed. The ECI-TOD coordinate frame is fixed in inertial space and is formed by a single negative rotation about the x_1 axis by the amount of the *obliquity*, or about 23.43928 deg for the year 2000 (Ref. 5, p. K6). This rotation simply aligns the $\bar{x}_3$ axis to coincide with Earth's axis of rotation and defines the ECI-TOD coordinate frame relative to the ecliptic plane. Although the ECI frame does have some angular motion from epoch to epoch, the effects of this motion are very small, especially over the brief time intervals used for this application. The precession and nutations of the Earth's axes are described in Eqs. (1.70–1.72) in Sec. 1.2.6.

Using these assumptions, we can write the transformation equations for the position, velocity, and acceleration components relative to the ECI frame as shown in Fig. 3.2:

$$\bar{x}^i = a^i_\alpha x^\alpha - a^i_\alpha X^\alpha \text{ for postion} \tag{3.44}$$

$$\bar{v}^i = a^i_\alpha v^\alpha - a^i_\alpha V^\alpha \text{ for velocity} \tag{3.45}$$

$$\dot{\bar{v}}^i = a^i_\alpha \dot{v}^\alpha - a^i_\alpha \dot{V}^\alpha \text{ for acceleration} \tag{3.46}$$

These equations are possible because the derivative of the transformation matrix a is assumed to be zero in this analysis. Notice the notation in Fig. 3.2 is intended to define the direction of the vectors, i.e., the subscript ES refers to the vector direction from the Earth to the sun, and EL refers to vectors from the Earth to the moon (lunar). For example, the components of $\boldsymbol{R}_{\text{SE}}$ are the X^α in Eq. (3.44).

The equations of motion for a space vehicle in inertial gravitational space are given by Eq. (3.14). The space vehicle acceleration in the ECI frame is given by Eq. (3.46) and must include the acceleration of the ECI frame itself, i.e., the $\dot{V}^\alpha$ components in Eq. (3.46). The acceleration of the Earth (the ECI frame) in the *ecliptic inertial coordinate frame* is as follows:

$$\dot{V}_i = \mu_S(|\boldsymbol{R}_{\text{ES}}|)^{-3} a_i^\alpha \bar{X}_\alpha + \mu_L(|\boldsymbol{R}_{\text{EL}}|)^{-3} a_i^\alpha \bar{X}_{L_\alpha} \tag{3.47}$$

Notice that $X_i = -a_i^\alpha \bar{X}_\alpha$, where the $\bar{X}_\alpha$ are the ECI components of the vector from the *Earth to the sun*, and $X_{L_i} = a_i^\alpha \bar{X}_{L_\alpha}$ are components of the vector from the *Earth to the moon*. These equations are written with these sign conventions in anticipation of the directions of the vectors to the sun and moon from the computational algorithms given in Ref. 5. Transforming Eq. (3.47) into the ECI coordinate frame, we have the acceleration of the Earth in ECI coordinates, the second term on the right-hand side of Eq. (3.46):

$$\dot{\bar{V}}_i = \mu_S(|\bar{\boldsymbol{R}}_{\text{ES}}|)^{-3} \bar{X}_i + \mu_L(|\bar{\boldsymbol{R}}_{\text{EL}}|)^{-3} \bar{X}_{L_i} \tag{3.48}$$

The μ_S and μ_L are the gravitational constants for the sun and moon, respectively. These constants are usually expressed in terms of a ratio times the Earth's gravitational constant and are given by Ref. 5 (p. K6), as follows:

$$\mu_S = k_s \mu_E, \text{ where } k_s = 332946.038$$

$$\mu_L = k_l \mu_E, \text{ where } k_l = 0.012300034$$

In a similar manner, we can write the acceleration of the space vehicle in the ECI frame; the first term on the right-hand side of Eq. (3.46) becomes

$$a_{\alpha i} \dot{v}^\alpha = \frac{\bar{f}_i}{m} - \mu_E(|\bar{\boldsymbol{R}}_{EV}|)^{-3} \bar{x}_i + \mu_S(|\bar{\boldsymbol{R}}_{VS}|)^{-3} \bar{x}_{VS_i} + \mu_L(|\bar{\boldsymbol{R}}_{VL}|)^{-3} \bar{x}_{VL_i} \tag{3.49}$$

where m is the mass of the space vehicle. The vectors $\boldsymbol{R}_{EV}$, $\boldsymbol{R}_{VS}$ and $\boldsymbol{R}_{VL}$ are vectors, Earth to vehicle, vehicle to sun, and vehicle to moon, respectively, as shown in Figure 3.2. From the equations given in Fig. 3.2, notice that

$$\bar{x}_{VS_i} = \bar{X}_i - \bar{x}_i$$

and

$$\bar{x}_{VL_i} = \bar{X}_{L_i} - \bar{x}_i \tag{3.50}$$

Finally, the acceleration of the space vehicle in the ECI coordinate frame becomes Eq. (3.49) minus Eq. (3.48), as written in Eq. (3.46):

$$\dot{\bar{v}}_i = a_{\alpha i} \dot{v}^\alpha - \dot{\bar{V}}_i \tag{3.51}$$

Numerical integrals of this equation provide the ECI position and velocity ephemerides for the space vehicle. Lunar relative ephemerides are then given by the second equation of Eq. (3.50). Earth-lunar families of trajectories can be selected by defining the initial conditions at the starting epoch based on the following criteria:

1) The space vehicle's initial parking orbit around the Earth is planned to be in the Earth-lunar orbital plane; this defines Ω, the right ascension of the orbit, and i, the inclination of the orbit. The altitude of this orbit will be the perigee of the Earth-lunar transfer orbit.

2) The line of apsides of the Earth-lunar transfer orbit is planned to pass in front of the moon, thus defining ω, the argument of perigee of the orbit. The line of apsides is adjusted so that the lunar relative velocity will define a hyperbolic lunar-Earth transfer orbit.

3) The lunar-Earth transfer orbit is planned for a safe return to Earth in the event that the braking rocket engine burn to the lunar parking orbit is not performed.

4) The lunar-Earth transfer orbit is designed for a) a safe reentry into the Earth's atmosphere with a landing at the desired target point on the Earth or b) to place the space vehicle into the desired parking orbit around the Earth for a later de-orbit to landing maneuver.

3.2 Rigid Body Dynamics

3.2.1 Equations of Motion for a Rigid Body in Gravitational Space

Using the fundamental total energy Eq. (3.12) and the same assumptions as in the potential energy analysis of Sec. 3.1.4, we can now examine the motion of the rigid body near a large planetary body, such as the Earth. Figure 3.3 will help to visualize the aerospace vehicle as a system of mass particles and define the coordinates used.

In this analysis the planetary mass is considered stationary at the center of the inertial frame, i.e., both $\boldsymbol{r}^{\gamma}$ and $\boldsymbol{v}^{\gamma}$ remain equal to zero, and the rate of change of energy of the system of mass particles in Eq. (3.13) reduces to

$$f^{\rho} \cdot \boldsymbol{v}_{\rho} = m^{\rho}(\dot{\boldsymbol{v}}_{\rho} \cdot \boldsymbol{v}_{\rho}) + \mu m^{\rho}(|\boldsymbol{r}_{\rho}|)^{-3}(\boldsymbol{r}_{\rho} \cdot \boldsymbol{v}_{\rho}) \tag{3.52}$$

where the summations on ρ will represent the space vehicle in Fig. 3.3. The inertial vector $\boldsymbol{R}$ is to the center of mass of the space vehicle. To simplify the analysis, let

$$g_{\rho i} = \mu m_{\rho}(|\boldsymbol{r}_{\rho}|)^{-3} x_{\rho i} \tag{3.53}$$

be the *gravitational force* felt by each mass particle of the space vehicle. Notice Eq. (3.53) is written for the ρth mass particle and we *do not sum* on ρ at this point in the analysis. Using Eq. (3.25) for the position components, Eq. (3.28) for velocity components, and Eq. (1.233) for the acceleration components in Eq. (3.52), we

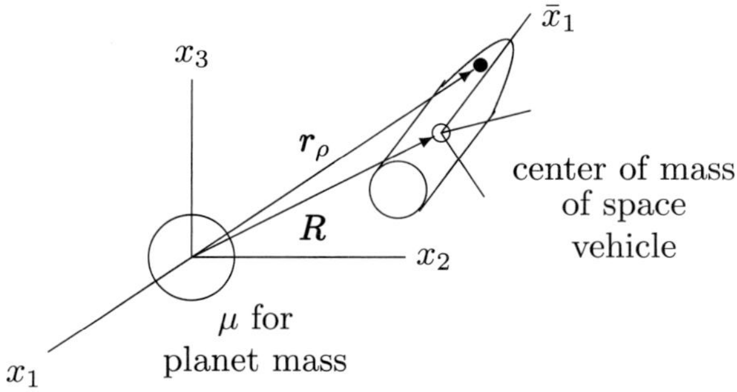

Fig. 3.3 The space vehicle as a system of mass particles.

begin with the following intermediate relationships:

$$\sum_{i=1}^{3}(a_i^\alpha \bar{f}_{\rho\alpha})(a_i^\beta \bar{\mathcal{X}}_\beta^{\rho\gamma}\bar{\omega}_\gamma + V_i)$$

$$= \sum_{i=1}^{3} m_\rho(a_i^\alpha \bar{a}_\alpha^\rho + \dot{V}_i)(a_i^\beta \bar{\mathcal{X}}_\beta^{\rho\gamma}\bar{\omega}_\gamma + V_i) + \sum_{i=1}^{3}(a_i^\alpha \bar{g}_{\rho\alpha})(a_i^\beta \bar{\mathcal{X}}_\beta^{\rho\gamma}\bar{\omega}_\gamma + V_i) \tag{3.54}$$

where the

$$f_i^\rho = a_i^\alpha \bar{f}_\alpha^\rho \text{ or } f_{\rho i} = a_i^\alpha \bar{f}_{\rho\alpha} \tag{3.55}$$

and likewise,

$$g_i^\rho = a_i^\alpha \bar{g}_\alpha^\rho \text{ or } g_{\rho i} = a_i^\alpha \bar{g}_{\rho\alpha} \tag{3.56}$$

are the components of the contact forces and *gravitational forces*, respectively, acting on each mass particle. The sum of all contact forces acting on the space vehicle in inertial coordinates becomes

$$F_i = \sum_{\rho=1}^{n} f_i^\rho \tag{3.57}$$

In a similar manner, let

$$\mathcal{G}_i = \sum_{\rho=1}^{n} g_i^\rho \tag{3.58}$$

be the sums of all the *gravitational forces* acting on each mass particle, likewise in inertial coordinates.

The summation signs in Eq. (3.54) are necessary to carry out the vector dot products. Again, the notation $\bar{f}_{\rho i}$ is used to preserve the implied summation of the $\bar{f}_i$ components of force acting on each mass particle of the space vehicle. Expanding Eq. (3.54) and using the center of mass conditions as described in Sec. 3.1.1, we have the second intermediate relationships:

$$\sum_{i=1}^{3}(\bar{f}_{\rho i}\bar{\mathcal{X}}_i^{\rho\gamma}\bar{\omega}_\gamma + F_iV_i)$$

$$= \sum_{i=1}^{3}(m_\rho\bar{a}_i^\rho\bar{\mathcal{X}}_i^{\rho\gamma}\bar{\omega}_\gamma + M\dot{V}_iV_i) + \sum_{i=1}^{3}(\bar{g}_{\rho_i}\bar{\mathcal{X}}_i^{\rho\gamma}\bar{\omega}_\gamma + \mathcal{G}_iV_i) \qquad (3.59)$$

where $\mathcal{G}_i$ is the total gravitation force vector components acting on the space vehicle, as given in Eq. (3.58).

The differential equations given in the intermediate equations of Eq. (3.59) separate into the equations of motion for the space vehicle. This derivation in classical mechanics, here using Cartesian tensor analysis, demonstrates that there is no cross coupling of gravitational energy between rotations of the *rigid body* and its translational motion, i.e., the rotational dynamics separate out from the translational dynamics. This is not true if the body is not rigid but is deformed and acquires energy from the resulting gravitational forces. However, in this analysis (the rigid body case), the translational equations of motion for the center of mass of the space vehicle, which constitute Newton's laws of motion and Newton's law of gravitation, are formed by equating the products of *Vi*, and we have

$$F_i - \mathcal{G}_i = M\dot{V}_i \qquad (3.60)$$

The rotational equations of motion about the center of mass are formed by equating products of $\bar{\omega}_\gamma$ in Eq. (3.59), and we have the intermediate equation:

$$\sum_{i=1}^{3}\bar{f}_{\rho i}\bar{\mathcal{X}}_i^{\rho\gamma} - \sum_{i=3}^{3}\bar{g}_{\rho i}\bar{\mathcal{X}}_i^{\rho\gamma} = \sum_{i=1}^{3}m_\rho\bar{a}_i^\rho\bar{\mathcal{X}}_i^{\rho\gamma} \qquad (3.61)$$

where there will be three sets of these equations as $\gamma = 1, 2$, and 3, respectively. Detailed expansion and reduction of these equations are the subjects of the next sections of the text.

3.2.2 Center of Gravity of a Rigid Body

The translational and rotational dynamics of a rigid body in gravitational space are described by the motion of the center of mass of the body. When a gravitational gradient exists, the sum of the gravitational forces on each particle of mass of the body may not act precisely through the center of mass of the body. Similar to the center of mass concepts of Sec. 3.1.1, we can write:

$$\mathcal{G}_i = \mu m_\rho(|\boldsymbol{r}^\rho|)^{-3}x_i^\rho = \mu M(|\boldsymbol{R}_{\text{cg}}|)^{-3}(X_i + \Delta X_{\text{cg}_i}) \qquad (3.62)$$

where the $\boldsymbol{R}_{\text{cg}}$ represents the inertial vector to the *center of gravity* of the rigid body with vector components referring to the center of mass with the vector components

$X_i + \Delta X_{cg_i}$. Although the center of gravity offset from the center of mass does not find many applications in rigid body dynamics, it does cause the gravity gradient torques to effect the rotational motion of space vehicles. In this section we will derive the estimates for the center of gravity of a space vehicle, however small it may be, and in Sec. 3.2.4 we will derive the approximating expression for the gravity gradient torques without the necessity of determining the space vehicle's center of gravity.

Using the radius vector magnitude given in Eq. (1.69) and selecting only the first two terms of the binomial expansion, we have the approximation

$$|\boldsymbol{r}^{\rho}|^{-3} \cong R^{-3} - 3R^{-5}(\tilde{\boldsymbol{x}}^{\rho} \cdot \boldsymbol{R}) \tag{3.63}$$

Placing this into the left-hand sides of Eq. (3.62), we have the matrix equation

$$\begin{pmatrix} \mathcal{G}_1 \\ \mathcal{G}_2 \\ \mathcal{G}_3 \end{pmatrix} = \mu M R^{-3} \begin{pmatrix} X_1 \\ X_2 \\ X_3 \end{pmatrix} + 3\mu R^{-4} (\mathcal{I}') \begin{pmatrix} u_1 \\ u_2 \\ u_3 \end{pmatrix} \tag{3.64}$$

Notice that the equation is left referencing the inertial coordinate frame, and

$$(\mathcal{I}') = \begin{pmatrix} I_{11} - (1/2)T_R & I_{12} & I_{13} \\ I_{21} & I_{22} - (1/2)T_R & I_{23} \\ I_{31} & I_{32} & I_{33} - (1/2)T_R \end{pmatrix} \tag{3.65}$$

where $T_R = I_{11} + I_{22} + I_{33}$, is called the *trace of a matrix*, and in this case, the summations form the trace of the inertial moment-of-inertia matrix. The u_i are the components of the unit vector in the radius vector direction, i.e., $u_i = X_i/R$, and is also used to reduce the term to the order of R^{-4}.

In a similar manner, the right-hand side of Eq. (3.62) reduces to the approximating equation

$$\begin{pmatrix} \mathcal{G}_1 \\ \mathcal{G}_2 \\ \mathcal{G}_3 \end{pmatrix} = \mu M R^{-3} \begin{pmatrix} X_1 \\ X_2 \\ X_3 \end{pmatrix} + 3\mu M R^{-3} (\mathcal{X}') \begin{pmatrix} \Delta X_{cg_1} \\ \Delta X_{cg_2} \\ \Delta X_{cg_3} \end{pmatrix} \tag{3.66}$$

where the matrix

$$(\mathcal{X}') = \begin{pmatrix} (1/3) - u_1 u_1 & -u_1 u_2 & -u_1 u_3 \\ -u_2 u_1 & (1/3) - u_2 u_2 & -u_2 u_3 \\ -u_3 u_1 & -u_3 u_2 & (1/3) - u_3 u_3 \end{pmatrix} \tag{3.67}$$

and again the u_i are the components of the unit vector in the $\boldsymbol{R}$ direction. Placing Eqs. (3.66) and (3.64) equal to each another and solving for the ΔX_{cg_i} vector, we have the approximation for the center of gravity offset from the center of mass of the space vehicle, which is given here in inertial coordinates:

$$\begin{pmatrix} \Delta X_{cg_1} \\ \Delta X_{cg_2} \\ \Delta X_{cg_3} \end{pmatrix} = \left(\frac{1}{MR} \right) (\mathcal{X}')^{-1} (\mathcal{I}') \begin{pmatrix} u_1 \\ u_2 \\ u_3 \end{pmatrix} \tag{3.68}$$

The center of gravity offset is a function of the space vehicle's size, mass, and angular orientation with respect to the radius vector direction. Generally, for a space vehicle weighing about 70,000 lb and having a reference length of 120 ft, the center of gravity is offset at times only by about 0.70×10^{-3} in. from the center of mass. However, at orbital altitudes, the total gravitational force will be less than its reference 1-g weight and will act through this point and will produce rotational torques on the space vehicle.

3.2.3 Euler's Rotational Equations of Motion for a Rigid Body

For the analysis presented in this section, we begin with Eq. (3.61) and purposely omit the second terms on the left-hand side of the equation. This second term will be discussed in detail in the next section.

Using the $\bar{\mathcal{X}}_i^{\rho\gamma}$ matrix components given in Eq. (3.19), we can expand the first term on the left-hand side of Eq. (3.61) for $\gamma = 1, 2$, and 3, as follows:

$$\begin{aligned} \bar{x}_2^{\rho}\bar{f}_{\rho 3} - \bar{x}_3^{\rho}\bar{f}_{\rho 2} &= \bar{N}_1 \\ \bar{x}_3^{\rho}\bar{f}_{\rho 1} - \bar{x}_1^{\rho}\bar{f}_{\rho 3} &= \bar{N}_2 \\ \bar{x}_1^{\rho}\bar{f}_{\rho 2} - \bar{x}_2^{\rho}\bar{f}_{\rho 1} &= \bar{N}_3 \end{aligned} \tag{3.69}$$

The $\bar{N}_i$ are the vector components of the sum of all applied torques acting on the space vehicle in body axis coordinates.

To expand the term on the right-hand side of Eq. (3.61), it makes the analysis simpler to use a slightly different but equivalent expression for the $\bar{a}_i^{\rho}$ components as presented in Eq. (1.232), namely,

$$\bar{a}_i^{\rho} = \bar{W}_i^{\sigma}\bar{\mathcal{X}}_{\sigma}^{\rho\alpha}\bar{\omega}_{\alpha} + \bar{\mathcal{X}}_i^{\rho\beta}\dot{\bar{\omega}}_{\beta} \tag{3.70}$$

realizing again that all of the $\bar{v}_i^{\rho}$ and $\dot{\bar{v}}_i^{\rho}$ components will be zero for the rigid body case. Using this expression in the right-hand side of Eq. (3.61), we have the intermediate result:

$$\sum_{i=1}^{3} m_{\rho}\bar{a}_i^{\rho}\bar{\mathcal{X}}_i^{\rho\gamma}$$
$$= \sum_{i=1}^{3} m_{\rho}\bar{W}_i^{\sigma}\bar{\mathcal{X}}_{\sigma}^{\rho\alpha}\bar{\mathcal{X}}_i^{\rho\gamma}\bar{\omega}_{\alpha} + \sum_{i=1}^{3} m_{\rho}\bar{\mathcal{X}}_i^{\rho\beta}\bar{\mathcal{X}}_i^{\rho\gamma}\dot{\bar{\omega}}_{\beta} \tag{3.71}$$

The first term on the right-hand side in Eq. (3.71) forms tensor-like sums not like a matrix multiply, however, it expands to equivalent matrix operations. The second summation on the right-hand side of this equation forms the moment-of-inertia tensor as given in Eq. (3.21). Using the matrix equivalents for the sums formed in the first expression of Eq. (3.71), the inertia tensor, and equating to the applied torques in Eq. (3.69), we have Euler's rotational equations of motion for a rigid body:

$$\bar{N}_i = \bar{W}_i^{\alpha}\bar{I}_{\alpha}^{\beta}\bar{\omega}_{\beta} + \bar{I}_i^{\gamma}\dot{\bar{\omega}}_{\gamma} \tag{3.72}$$

The second terms on the right-hand side of this equation are the gyroscopic inertia torques, which constitutes Newton's law of inertia for a rotating rigid body.

Euler's rotational equations of motion can also be derived using the tensor qualities of the angular momentum vector and Newton's law for the rate of change of the angular momentum vector as given in Eq. (2.27):

$$N_i = \frac{\mathrm{d}L_i}{\mathrm{d}t} \tag{3.73}$$

Using the tensor relationships for the angular momentum as given in Eq. (3.32), we have

$$a_i^\alpha \bar{N}_\alpha = \frac{\mathrm{d}(a_i^\alpha \bar{I}_\alpha^\beta \bar{\omega}_\beta)}{\mathrm{d}t} \tag{3.74}$$

Differentiating the right-hand side of Eq. (3.74) and remembering that the inertia tensor is a constant for a rigid body, we can write

$$a_i^\alpha \bar{N}_\alpha = a_i^\alpha (\bar{W}_\alpha^\beta \bar{I}_\beta^\gamma \bar{\omega}_\gamma + \bar{I}_\alpha^\beta \dot{\bar{\omega}}_\beta)$$

and

$$\bar{N}_i = \bar{W}_i^\beta \bar{I}_\beta^\gamma \bar{\omega}_\gamma + \bar{I}_i^\beta \dot{\bar{\omega}}_\beta \tag{3.75}$$

Again, we have Euler's rotational equations of motion for a rotating rigid body.

When rotating subbodies are part of the rigid body rotational analysis, their effects can be included by starting with Eq. (3.35) for the angular momentum in Eq. (3.74). Again differentiating, as in Eq. (3.74), we have Euler's rotational equations of motion for a rigid body, for the general case, including the spinning subbody:

$$\bar{N}_i = \bar{W}_i^\alpha (\bar{I}_\alpha^\beta \bar{\omega}_\beta + b_\alpha^\beta \bar{\bar{I}}_\beta^\gamma \bar{\bar{\omega}}_\gamma) + \bar{I}_i^\gamma \dot{\bar{\omega}}_\gamma + b_i^\alpha (\bar{\bar{W}}_\alpha^\beta \bar{\bar{I}}_\beta^\gamma \bar{\bar{\omega}}_\gamma + \bar{\bar{I}}_\alpha^\beta \dot{\bar{\bar{\omega}}}_\beta) \tag{3.76}$$

Notice that the vector components $\bar{\bar{W}}_\alpha^\beta \bar{\bar{I}}_\beta^\gamma \bar{\bar{\omega}}_\gamma$ will be zero for the *balanced spinning subbody*. Also notice that for a spinning subbody that has a constant rotation rate, i.e., $\dot{\bar{\bar{\omega}}}_\beta = 0$, then Eq. (3.76) reduces to

$$\bar{N}_i = \bar{W}_i^\alpha (\bar{I}_\alpha^\beta \bar{\omega}_\beta + b_\alpha^\beta \bar{\bar{I}}_\beta^\gamma \bar{\bar{\omega}}_\gamma) + \bar{I}_i^\alpha \dot{\bar{\omega}}_\alpha \tag{3.77}$$

3.2.4 Rigid Body Gravity Gradient Torques

We now examine the similarities between the terms on the left-hand side of Eq. (3.61). Notice that this equation is in the body axis coordinate reference frame. However, the *gravitational force* felt by the ρth mass particle, $g_{\rho i}$ as shown in Eq. (3.53), is written in the inertial reference frame. In Eq. (3.61), each of the inertial gravity force components have been transformed into the body axis coordinate frame by

$$\bar{g}_{\rho i} = \mu m_\rho (|\boldsymbol{r}_\rho|)^{-3} (\bar{x}_{\rho i} + \bar{X}_i) \tag{3.78}$$

where the $\bar{X}_i$ are the components of the radius vector to the center of mass of the space vehicle, which have also been transformed into the body axis coordinate frame. The torque equations, similar to Eqs. (3.69), are formed with the gravitational force and are summed on all mass particles making up the space vehicle:

$$\begin{aligned}
\bar{x}_3^\rho \bar{g}_{\rho 2} - \bar{x}_2^\rho \bar{g}_{\rho 3} &= \bar{N}_{g_1} \\
\bar{x}_1^\rho \bar{g}_{\rho 3} - \bar{x}_3^\rho \bar{g}_{\rho 1} &= \bar{N}_{g_2} \\
\bar{x}_2^\rho \bar{g}_{\rho 1} - \bar{x}_1^\rho \bar{g}_{\rho 2} &= \bar{N}_{g_3}
\end{aligned} \tag{3.79}$$

Approximations for these equations will form the gravitational torques, as shown in Ref. 43, which will affect the rotational dynamics of the space vehicle.

Using the same approximation as in Eq. (3.63) and realizing that the vector dot product is invariant during coordinate transformations, we have

$$|\boldsymbol{r}^\rho|^{-3} \cong R^{-3} - 3R^{-5}(\bar{\boldsymbol{r}}^\rho \cdot \bar{\boldsymbol{R}}) \tag{3.80}$$

where the $\bar{X}_i$ are the same as in the preceding and are the components of the vector $\bar{\boldsymbol{R}}$. Using this approximation in Eq. (3.78) and substituting into Eqs. (3.79), we have the desired relationship approximating the gravitational torques on the space vehicle:

$$\bar{\boldsymbol{N}}_g = \begin{pmatrix} \bar{N}_{g1} \\ \bar{N}_{g2} \\ \bar{N}_{g3} \end{pmatrix} = 3\mu R^{-3} \begin{pmatrix} 0 & -\bar{u}_3 & \bar{u}_2 \\ \bar{u}_3 & 0 & -\bar{u}_1 \\ -\bar{u}_2 & \bar{u}_1 & 0 \end{pmatrix} \begin{pmatrix} \bar{I}_{11} & \bar{I}_{12} & \bar{I}_{13} \\ \bar{I}_{21} & \bar{I}_{22} & \bar{I}_{23} \\ \bar{I}_{31} & \bar{I}_{32} & \bar{I}_{33} \end{pmatrix} \begin{pmatrix} \bar{u}_1 \\ \bar{u}_2 \\ \bar{u}_3 \end{pmatrix} \tag{3.81}$$

where the $\bar{u}_i$ are the unit vector components of the vector $\bar{\boldsymbol{R}}$. Also notice indeed that the summations on ρ for all mass particles form the body-axis-moment-of-inertia tensor. This property of the derivation makes the application of this analysis especially useful to the aerospace engineer when working problems involving on-orbit space vehicle attitude stability and control.

3.2.5 Aerospace Applications: Numerical Solution of Euler's Rotational Equations of Motion for a Rigid Body

In most cases, the solution of Euler's rotational equations of motion, Eq. (3.72), for the rotational motion of a rigid vehicle can only be solved numerically using digital computer systems. The six-DOF terminology refers to coordinates that are free to move in the $\bar{x}_1$, $\bar{x}_2$, and $\bar{x}_3$ directions as well as being free to rotate about each coordinate axis, i.e., with varying $\bar{\omega}_1$, $\bar{\omega}_2$, and $\bar{\omega}_3$ body axis rotation rates. In Sec. 1.4, the Hamilton quaternion is shown to be excellent parameters to describe the rotational dynamics of rotating bodies. By using the first integral of Euler's rotational equations of motion and integrating the quaternion rates from Eq. (1.217),[44] which relates body axis rotation rates to the quaternion rates, the dynamic body axis attitude quaternion results. The body axis attitude Euler angles are then computed from the attitude matrix given by Eq. (1.111). This process provides a very accurate and continuous solution for the vehicle attitude.

Note in this application we are describing only the rotational motion solution for the attitude of the vehicle and do not include the three DOF for the translational motion. The integral list for the six-DOF solution for vehicle motion is summarized in Appendix C.3.

The body axis torques $\bar{N}_i$ on the left-hand side will represent the sum of all torques acting on the vehicle and will depend on the specific aerospace application. These torques may result from the following forces: 1) engine thrust not vectored through the center of mass; 2) aerodynamic pressure forces; 3) control torques, usually using control effector sources from 1) and 2); 4) gravity gradients; and 5) solar radiation pressure forces.

The solution steps involved are the following:

Step 1: Using the body axis torques from the application software, Eq. (3.72) computes the body axis rotation rate derivatives:

$$\dot{\bar{\omega}}_i = \bar{J}_i^{\alpha}\left(\bar{N}_\alpha - \bar{W}_\alpha^{\beta}\bar{I}_\beta^{\gamma}\bar{\omega}_\gamma\right) \tag{3.82}$$

where the $\bar{J}_i^{\alpha}$ are the components of the inverse of the body-axis-moment-of-inertia matrix, and in matrix notation is simply $(\bar{I})^{-1}$.

Step 2: The quaternion rates are computed from Eq. (1.217), repeated here for clarity:

$$\begin{pmatrix} \dot{q}_1 \\ \dot{q}_2 \\ \dot{q}_3 \\ \dot{q}_4 \end{pmatrix} = \frac{1}{2}\begin{pmatrix} 0 & -\bar{\omega}_1 & -\bar{\omega}_2 & -\bar{\omega}_3 \\ \bar{\omega}_1 & 0 & \bar{\omega}_3 & -\bar{\omega}_2 \\ \bar{\omega}_2 & -\bar{\omega}_3 & 0 & \bar{\omega}_1 \\ \bar{\omega}_3 & \bar{\omega}_2 & -\bar{\omega}_1 & 0 \end{pmatrix}\begin{pmatrix} q_1 \\ q_2 \\ q_3 \\ q_4 \end{pmatrix} \tag{3.83}$$

Step 3: An array vector of the seven integrable parameters from Steps 1 and 2 is defined as follows;

$x_i(t)$	$\dot{x}_i(t)$
$\bar{\omega}_1$	$\dot{\bar{\omega}}_1$
$\bar{\omega}_2$	$\dot{\bar{\omega}}_2$
$\bar{\omega}_3$	$\dot{\bar{\omega}}_3$
q_1	$\dot{q}_1$
q_2	$\dot{q}_2$
q_3	$\dot{q}_3$
q_4	$\dot{q}_4$

Step 4: These parameters become the input to first-order differential equation solution methods, as suggested and described in Appendix D. These methods are generally designed to integrate n simultaneous differential equations over a single time step Δt, with a functional evaluation after each step, and is done repetitively until the solution is completed to the desired boundary condition. The ending

boundary condition may be simply, for instance, when the desired stop time is reached. As shown in Appendix D.1, the single time step of the solution is

$$x_i(t + \Delta t) = x_i(t) + \int_t^{t+\Delta t} \dot{x}_i(t)\,\mathrm{d}t \qquad \text{for} \quad i = 1, 7 \tag{3.84}$$

Step 5: At any time during the solution process, Eq. (1.111) is used to form the body-axis-to-inertial transformation matrix, $a_i^\alpha = a_i^\alpha(q_\beta)$, and the required Euler angle set, as given in Appendix A, is extracted from the transformation matrix, (a) $\rightarrow \theta_1$, θ_2, and θ_3, defining the dynamic attitude of the vehicle.

Conclusion

In this chapter I have presented the reader with the fundamental mathematical concepts using N point masses to describe an N-body mass system. By writing the total energy relationship of the mass system as the sum of the kinetic energy and gravitational potential energy of each of the point masses, I have defined two large areas of study; 1) the N-body point mass system where the motion of each point mass is not constrained by a physical structure, as in the computation of N-body planetary motion, and; 2) N-body mass points which are bound in a physical structure, as in the dynamics of the rigid body.

I have tried to emphasize the importance of the basic center-of-mass relationships that are used in the development of the fundamental rigid body equations for energy, the moment of inertia tensor, and angular momentum. With these concepts in mind, I feel that the reader can better understand Euler's rotational equations of motion for a rigid body.

I define the differences between the center-of-mass and the center-of-gravity of a rigid body. With this concept the reader can visualize the gravitational torques acting on the space vehicle in orbit.

As an aerospace application, I have offered a numerical solution method for Euler's rotational equations of motion for a rigid body. The solution to this equation is required in many aerospace motion simulations and in general must be done numerically because of the complicated body-axis force, moment functions, and the fact that the moment of inertia tensors do not take the diagonal matrix form.

4
Flight Vehicle Motion

Introduction

This chapter is devoted to the modeling of actual gravitational accelerations and the forces and moments acting on the flight vehicle to simulate its real world motion. For example, the spherical gravitational field of the point mass, as discussed in the previous chapters, becomes the nonspherical field approximated by using the Legendre polynomial expansion to simulate the actual mass distribution of an astronomical body. The Two-Body Problem of astrodynamics becomes the Perturbed Two-Body Problem, where closed-form Keplerian motion solutions no longer describe the real motion of the space vehicle. Hence, motion solutions must be approximated by using perturbation analysis methods or the equations of motion must be numerically integrated using Cowell's method.

As a refresher for the reader, I have included subsections covering some of the important concepts of atmospheric modeling and also a brief description of the correspondence between aeronautical nomenclature and the variable names used in the text.

In the late 1960s, large scale digital computer system resources became availiable to the aerospace engineer. As a result a very exciting age began for the aerodynamist. Finite element aerodynamic modeling of the velocity flow fields over the flight vehicle surfaces are now possible. Using potential flow and fluid dynamic modeling the pressure coefficients on the vehicle surfaces can be estimated. This effectively simulates the actual airflows over the real flight vehicle. Both subsonic and supersonic airfoil sections, airplane wings, and indeed entire airplanes have been designed and tested long before the actual flight vehicle is test flown.

As mentioned above, the Keplerian orbital elements are no longer constants, but become the osculating orbital elements along the actual orbital path. A set of constant orbital elements, that represents the real orbit, can be defined at a specific epoch. These orbital elements are called the mean orbital elements and form the basis for viable methods to propagate and approximate the orbital motion of the real space vehicle.

The reader must realize that specific 6 degree-of-freedom (6-DOF) motion solutions are not discussed in the text. The reason is that when simulating actual flight vehicle motion, the specific end-boundary conditions are functions of each stepwise solution to the equations of motion. When simulating the effects of closed-loop controlled flight, these equations of motion are only described to be stepwise continuous. Therefore, over multiple time steps the equations of motion might change and are discontinuous because of the GNC control effector responses required to fly the vehicle in the desired flight envelope.

As an aerospace application example of this chapter, I describe a Flight Simulation Program for the solution of the 6-DOF motion of a typical flight vehicle. This example ties together many topics discussed in this chapter and shows the computer system logic environment required to fly the vehicle over each of the flight segements of its intended flight envelope. For instance, a rocket vehicle ascent to orbit might involve a series of contiguous segements, each requiring the stepwise solutions of different differential equations of motion. These segements for a launch vehicle might be: the vertical rise to clear the launch tower, roll to the launch azimuth, pitch to stay within the structural dynamic pressure limitations, rocket assembly staging, constant pitch-over rate steering, and finally, the controlled gravity turn until main engine cutoff and orbital injection.

4.1 Modeling Gravitational Environments for Aerospace Vehicles

4.1.1 *Nonspherical Gravitational Potentials*

The potential energy due to the gravitational field of an astronomical body is a function of the mass distribution of the body itself. When a flight vehicle is near a large astronomical mass, such as the Earth, equations of motion such as Eqs. (2.35) and (2.45) must be rewritten to include the gravitational effects of the non-homogeneous distribution of mass within the gravitating body.

The total energy of a system of mass particles at some point near a large planet, such as the Earth, can be defined by Eq. (3.12). For this analysis, we locate an inertial coordinate center at the center of mass of the planet, and in this case let the planet be represented by a system of ρ mass particles. Unlike the aerospace vehicle as depicted in Fig. 3.3, the vehicle here is simply represented by the point mass m_i. Summations indicated in Eq. (3.12) can be divided into the energy of the space vehicle and that of the planet, i.e., $E_T = E + E_P$, and we can write

$$E_T = (1/2)m_i v_i{}^2 - G m_i m^\rho (|\boldsymbol{r}_i - \boldsymbol{r}_\rho|)^{-1} + (1/2)\bar{\omega}^\alpha \bar{I}_\alpha^\beta \bar{\omega}_\beta - G m_\gamma m^\rho (|\boldsymbol{r}^\gamma - \boldsymbol{r}_\rho|)^{-1} \tag{4.1}$$

where the first term is the kinetic energy of the vehicle, and the second term is its potential energy relative to the mass of the planet. The third term is the kinetic energy of the planet due to its rotation in the inertial frame, which is given by Eq. (3.23). The kinetic energy of the planet due to translation is zero, since the planet's translational velocity is zero in this inertial coordinate frame. The body axis rotation rate components, the $\bar{\omega}_\alpha$, become simply (0, 0, ω_e), where for this case ω_e is the *sidereal rotation rate* of the Earth as described in Sec. 2.3.2. The $\bar{I}_\alpha^\beta$ are the components of the moment-of-inertia tensor of the planet. The fourth term is the potential energy of all mass particles of the planet, where the first sums over the planet masses will not include the mass of the vehicle, m_i, or simply $\gamma \neq i$. The planet's kinetic energy of rotation is considered to be constant. The planet's potential energy summations are constants except for that due to the space vehicle's small mass m_i, which we can assume will have no effect on the planet itself. The total energy simply becomes $E = E_T - E_P$. The system of mass particles involving only the vehicle and its potential energy relationship with the planet can be written from Eq. (4.1) simply as

$$E = (1/2)m_i v_i^{\,2} - G m_i m^\rho (|\boldsymbol{r}_i - \boldsymbol{r}_\rho|)^{-1} \tag{4.2}$$

Figure 4.1 depicts the geometry for this analysis. Using this figure, the gravitational potential energy of the planet at the center of mass of the aerospace vehicle becomes

$$\mathcal{V} = -G m m^\rho (|\boldsymbol{R} - \boldsymbol{r}_\rho|)^{-1} \tag{4.3}$$

The subscript i is dropped with the understanding that it refers only to the mass and position vector (changed to $\boldsymbol{R}$) to the center of mass of the space vehicle. The calligraphic $\mathcal{V}$ is used for the potential energy to be consistent with the notation generally used. Here again, the indicated summation on ρ will be over all mass particles representing the astronomical body.

In a similar manner as in the analysis of Sec. 3.1.6, we can write

$$\mathcal{V} = -G m m^\rho (R^2 + r_\rho^2 - 2\boldsymbol{R} \cdot \boldsymbol{r}_\rho)^{-(1/2)} \tag{4.4}$$

However, in this case the magnitude of $\boldsymbol{r}_\rho$ can be large compared to the magnitude of $\boldsymbol{R}$, and $\boldsymbol{r}_\rho$ can have motion in the inertial coordinate frame, such as the rotating Earth. For non-homogeneous bodies, closed-form integrals of Eq. (4.4) are not possible; however, approximate solutions can be found. For example, the analysis first performed by James MacCullagh[45] in 1855 was as follows: expand Eq. (4.4) using Legendre polynomials (Ref. 29, pp. 272–274), and perform the summations on ρ mass particles representing the planet. MacCullagh's analysis gave us the first estimates for the gravity potential of a planetary body:

$$\frac{\mathcal{V}}{m} \approx -\frac{GM}{R} - \left(\frac{G}{2R^3}\right)(T_R - 3\bar{I}_{11}) + \cdots \tag{4.5}$$

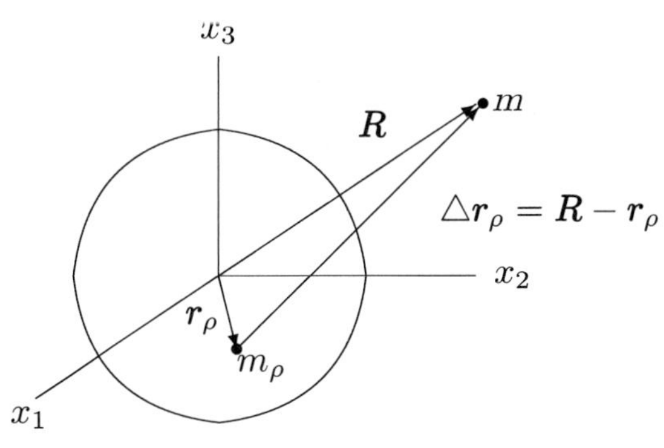

Fig. 4.1 The planet as a system of mass particles.

where T_R is the trace of the moment-of-inertia matrix of the planet, i.e., $T_R = I_{11} + I_{22} + I_{33}$. During the expansion of the polynomial terms (Ref. 8, pp. 509–513), the summation $m^\rho(r_\rho^2 \sin^2\theta_\rho)$ occurs. θ_ρ is the angle between the $\boldsymbol{R}$ and the $\boldsymbol{r}_\rho$ vectors. On close examination, this term forms the $\bar{I}_{11}$ element of the moment-of-inertia matrix about the radius vector $\boldsymbol{R}$. This element of the moment-of-inertia matrix can be computed from Eq. (3.26) and in matrix form is $\bar{I} = (c)^T\ (I)\ (c)$. The transformation matrix c is determined by a two-axis rotation of the ECI coordinate frame to the radius vector $\boldsymbol{R}$, which has the vector components (x_1, x_2, x_3). The transformation matrix becomes $c = Z(\theta)Y(-\phi)$, where $\theta = \tan^{-1}(x_2/x_1)$ and $\phi = \sin^{-1}(x_3/R)$.

Equation (4.5) is known as MacCullagh's formula, which offers an interesting relationship hinting to a connection between gravitation and the moment-of-inertia tensor. Notice further that the moment of inertia of a homogeneous sphere is $(2/5)Mr^2$, about any diameter, where r is the radius of the sphere. Representing the planet as a homogeneous sphere, MacCullagh's formula yields simply $-GM/R$. This is the identical results from the integrated solution for the gravity potential of a homogeneous sphere (Ref. 27, pp. 16–18).

The implied integral of Eq. (4.4) for a planet such as the Earth can be approximated very closely using the Legendre polynomial functions. We write the potential in the following form:

$$\mathcal{V} \approx -\frac{\mu m}{R}(1 + \mathcal{R}) \tag{4.6}$$

where $\mathcal{R}$ is the nonspherical perturbing potential. This potential is normally determined in body axis coordinates and here in *planet-fixed, planet-centered* spherical coordinates, where R is the radius distance to the space vehicle, θ is the planet-fixed longitude, and ϕ is the planet-centric latitude. Using these coordinates, the perturbing potential has the following form:

$$\begin{aligned}\mathcal{R} = &\sum_{n=2}^{N} C_{n,0}\left(\frac{R_e}{R}\right)^n P_n(\sin\phi)\\ &+ \sum_{n=2}^{N}\left(\frac{R_e}{R}\right)^n \sum_{m=1}^{n}\big(S_{n,m}\sin m\theta + C_{n,m}\cos m\theta\big)P_{n,m}(\sin\phi)\end{aligned} \tag{4.7}$$

R_e is the planet reference radius and is usually the equatorial radius. The Legendre polynomial function as given in Ref. 29 (p. 273) has the form as follows:

$$P_n(u) = \frac{1}{2^n n!}\frac{\mathrm{d}^n(u^2-1)^n}{\mathrm{d}u^n} \tag{4.8}$$

with the recursive relationship

$$P_{(n+1)}(u) = \frac{(2n+1)}{(n+1)}uP_n(u) - \frac{n}{(n+1)}P_{(n-1)}(u) \tag{4.9}$$

The notation here implies the following Legendre polynomials for $P_n(\sin\phi)$ in Eq. (4.7):

$$
\begin{aligned}
P_0(\sin\phi) &= 1 \\
P_1(\sin\phi) &= \sin\phi \\
P_2(\sin\phi) &= (1/2)(3\sin^2\phi - 1) \\
P_3(\sin\phi) &= (1/2)(5\sin^3\phi - 3\sin\phi) \\
P_4(\sin\phi) &= (1/8)(35\sin^4\phi - 30\sin^2\phi + 3)
\end{aligned}
\tag{4.10}
$$

The associated Legendre function as shown in Ref. 8 (pp. 513, 514) has the following form:

$$P_{n,m}(u) = (1-u^2)^{\frac{m}{2}} \frac{\mathrm{d}^m}{\mathrm{d}u^m} P_n(u) \tag{4.11}$$

The associated Legendre polynomials for $P_{n,m}(\sin\phi)$, like the conventional functions in Eqs. (4.10) are also given by Ref. 8 (p. 514). There are many excellent derivations of the expansion given in Eq. (4.7), such as those by Baker and Makemson in Ref. 31 (Sec. 5.2.2), Lear in Ref. 46, and Vallado in Ref. 8 (pp. 513–521).

The constants in Eq. (4.7) are called the gravity harmonic coefficients. Vallado presents an excellent discussion (Ref. 8, pp. 518–520) associating the gravity harmonics with the actual mass distribution of a planet. The gravity harmonics are named as follows:

1) $C_{n,0}$ are called the zonal coefficients and are sometimes written simply as $J_2, \ldots, J_N$. The zonal coefficients are functions of planet-centric latitude only.

2) $S_{n,m}$ and $C_{n,m}$, where $n \neq m$, are the tesseral coefficients and are functions of planet longitude.

3) $S_{n,n}$ and $C_{n,n}$ are the sectorial coefficients and are functions of both planet longitude and planet-centric latitude.

4.1.2 Nonspherical Gravitational Accelerations

The total energy relationship for the space vehicle, from Eq. (4.2), becomes

$$E \approx (1/2)mv^2 - \frac{\mu m}{R}(1 + \mathcal{R}) \tag{4.12}$$

Notice again that the gravitational potential function, as discussed in Eq. (4.7), is derived using planet-fixed, planet-centered spherical coordinates to approximate the non-homogeneous mass distribution of the planet. In the case of the Earth, the spherical coordinates are in the ECF system, the Earth-fixed, Earth-centered system. We can write Eq. (4.6) in functional form as follows:

$$\mathcal{V} = \mathcal{V}(R, \phi, \theta) \tag{4.13}$$

where the space vehicle is located relative to the planet-fixed coordinate frame. The planet itself has rotational energy that is part of the gravitational field, which must also be rotating in inertial space. The Earth-fixed longitude is related to the inertial longitude, as discussed in Sec. 2.3.2, by the following relationship:

$$\theta = \theta_i - GHA_o - \omega_e(t - t_o) \tag{4.14}$$

where

$\theta_i =$ the inertial longitude

$GHA_o =$ the Greenwich hour angle at time t_o

$\omega_e =$ the Earth's rotation rate magnitude

This makes Eq. (4.13) take the functional form

$$\mathcal{V} = \mathcal{V}(R, \phi, \theta(\theta_i, \omega_e, t)) \tag{4.15}$$

Hence, the parameter, Earth-fixed longitude becomes a function of time and has a time derivative, i.e., $\dot{\theta} = \dot{\theta}_i - \omega_e$. In a conservative system the time rate of change of the potential energy must be included in the derivation of the equations of motion as shown in Ref. 9 (p. 34) and Ref. 27 (pp. 137–140). Based on the analysis presented with Eq. (3.14) in Sec. 3.1.2, the time rate of change of the potential energy function given by Eq. (4.15) becomes

$$\frac{\mathrm{d}\mathcal{V}}{\mathrm{d}t} = \frac{\partial \mathcal{V}}{\partial R}\frac{\partial R}{\partial x_\alpha}\dot{x}^\alpha + \frac{\partial \mathcal{V}}{\partial \phi}\frac{\partial \phi}{\partial x_\alpha}\dot{x}^\alpha + \frac{\partial \mathcal{V}}{\partial \theta}\frac{\partial \theta}{\partial x_\alpha}\dot{x}^\alpha + \frac{\partial \mathcal{V}}{\partial \theta}\frac{\partial \theta}{\partial t} \tag{4.16}$$

From Eq. (4.14), we have

$$\frac{\partial \theta}{\partial t} = -\omega_e \tag{4.17}$$

The effects of the rotating gravitational field (Ref. 47, pp. 6-1 and 6-2) can be included in the equations of motion by multiplying the last term of Eq. (4.16) by $(\boldsymbol{v} \cdot \boldsymbol{v})/(v^2)$, or the factor

$$\frac{(\dot{x}_1^2 + \dot{x}_2^2 + \dot{x}_3^2)}{v^2} \tag{4.18}$$

This factor is equal to one and will not change the validity of the equation. Differentiating Eq. (4.12) with respect to time and using Eq. (4.16), we have the rate of change of energy, which can be written using vector notation as follows:

$$\frac{\mathrm{d}E}{\mathrm{d}t} = \boldsymbol{f} \cdot \boldsymbol{v} = m\dot{\boldsymbol{v}} \cdot \boldsymbol{v} - \nabla\mathcal{V} \cdot \boldsymbol{v} + \frac{\omega_e}{v^2}\frac{\partial \mathcal{V}}{\partial \theta}\boldsymbol{v} \cdot \boldsymbol{v} \tag{4.19}$$

Grouping the terms of like dot products, again using Eq. (4.16) and as described in Eq. (3.13), we have the Cartesian acceleration components in the inertial frame due to gravitation:

$$\ddot{x}_1 = \cos\theta_i\left(\cos\phi\frac{\partial\mathcal{V}}{\partial R} - \frac{1}{R}\sin\phi\frac{\partial\mathcal{V}}{\partial\phi}\right) - \left(\sin\theta_i + \frac{R\omega_e\dot{x}_1\cos\phi}{v^2}\right)\left(\frac{1}{R\cos\phi}\right)\frac{\partial\mathcal{V}}{\partial\theta}$$

$$\ddot{x}_2 = \sin\theta_i\left(\cos\phi\frac{\partial\mathcal{V}}{\partial R} - \frac{1}{R}\sin\phi\frac{\partial\mathcal{V}}{\partial\phi}\right) + \left(\cos\theta_i - \frac{R\omega_e\dot{x}_2\cos\phi}{v^2}\right)\left(\frac{1}{R\cos\phi}\right)\frac{\partial\mathcal{V}}{\partial\theta}$$

$$\ddot{x}_3 = \sin\phi\frac{\partial\mathcal{V}}{\partial R} + \frac{\cos\phi}{R}\frac{\partial\mathcal{V}}{\partial\phi} - \frac{\omega_e\dot{x}_3}{v^2}\frac{\partial\mathcal{V}}{\partial\theta} \qquad (4.20)$$

The effects of including the ω_e terms are small. The factor ω_e/v is about 3×10^{-9} for low Earth orbits. Hence, for motion in Earth orbit these terms are usually omitted. However, in a purely mathematical context, when these terms are omitted from the equations of motion during computer simulations of a flight vehicle in Earth orbit, the expected small oscillations in total energy result. Conversely, since Eqs. (4.20) were derived from Eq. (4.12) with $(\mathrm{d}E/\mathrm{d}t) = 0$, i.e., with no contact forces acting on the space vehicle, then the simulated total energy of the space vehicle is constant.

These simulation results, however, cannot verify the existence of the small accelerations resulting from time-dependent terms of the gravitational potential, as shown in Eq. (4.20). The difficulty is that the orbit determination processors, as described in Sec. 4.1.3, do not use the equations of motion as derived in Eq. (4.20). Hence, the *solved-for* gravity harmonic coefficients, as shown in Eq. (4.7), as determined may indeed include the real perturbing effects of the rotating gravity field of the planet. Therefore, when using these harmonic coefficients in the acceleration Eqs. (4.20) and as described in the following, the resulting small perturbations to the motion may be fictitious and will not simulate the real motion of the space vehicle. This problem, i.e., solving the two-body orbital equations of motion that include time-dependent gravitational potentials, is addressed in the methods of Bond and Gottlieb given in Ref. 48. These methods can be used to describe the perturbations to the space vehicle's motion when simulating the effects of large rotating gravity fields of astronomical bodies and are briefly outlined in Sec. 4.3.2.

The partial derivatives per unit mass from Lear (Ref. 46, pp. 2–6) in Eqs. (4.20) are formed from Eq. (4.7) and are as follows:

$$\frac{\partial\mathcal{V}}{\partial R} = \frac{\mu}{R^2}\left[1 + \sum_{n=2}^{N}(n+1)C_{n,0}\left(\frac{R_e}{R}\right)^n P_n(\sin\phi) + \sum_{n=2}^{N}(n+1)\left(\frac{R_e}{R}\right)^n\sum_{m=1}^{n}(S_{n,m}\sin m\theta + C_{n,m}\cos m\theta)P_{n,m}(\sin\phi)\right]$$

$$\frac{\partial\mathcal{V}}{\partial\phi} = -\frac{\mu}{R}\left[\sum_{n=2}^{N}C_{n,0}\left(\frac{R_e}{R}\right)^n\left(\frac{\partial P_n(\sin\phi)}{\partial\phi}\right) + \sum_{n=2}^{N}\left(\frac{R_e}{R}\right)^n\sum_{m=1}^{n}(S_{n,m}\sin m\theta + C_{n,m}\cos m\theta)\left(\frac{\partial P_{n,m}(\sin\phi)}{\partial\phi}\right)\right]$$

$$\frac{\partial \mathcal{V}}{\partial \theta} = -\left(\frac{\mu}{R}\right)\sum_{n=2}^{N}\left(\frac{R_e}{R}\right)^n \sum_{m=1}^{n} m(S_{n,m}\cos m\theta - C_{n,m}\sin m\theta)P_{n,m}(\sin\phi) \tag{4.21}$$

Continuing with the Lear formulation, the preceding partial derivatives are defined with primed terms as follows:

$$\begin{aligned}\frac{\partial P_n(\sin\phi)}{\partial \phi} &= \cos\phi P'_n(\sin\phi)\\ \frac{\partial P_{n,m}(\sin\phi)}{\partial \phi} &= P'_{n,m}(\sin\phi)\end{aligned} \tag{4.22}$$

and have the recursive relationships as follows:

$$\begin{aligned}P'_n &= \sin\phi P'_{n-1} + nP_{n-1}\\ \cos\phi P'_{n,m} &= -n\sin\phi(\sec\phi P_{n,m}) + (n+m)(\sec\phi P_{(n-1),m})\end{aligned} \tag{4.23}$$

Equations (4.21) and their application using computer algorithms are, in themselves, a study in the use of the associated Legendre functions. For a more detailed analysis of the gravitational acceleration equations and their application, see the additional work by Lear,[49] as well as the works of Spencer [50] and Gottlieb.[51]

4.1.3 Aerospace Applications: Orbit Determination Processors to Solve for Nonspherical Gravity Harmonic Coefficients

The zonal, tesseral, and sectorial gravity coefficients, as discussed in this section, can be determined statistically as *solved-for parameters* in orbit determination solution processes. The orbit determination processors can use least-squares analysis (Ref. 8, Secs. 10.2–10.5) to vary parameters in the differential equations of motion to fit the orbital paths to actual ground tracking data. Additional examples of orbit determination methods can be found in Ref. 42 (pp. 79–96) and Ref. 28 (pp. 394–400).

For a better understanding of the overall process, we start by using functional notation to represent the propagation methods for the solutions of the flight vehicle's equations of motion. The equations of motion could include the gravitational accelerations in Eqs. (4.19) and (4.20) to describe the motion of a space vehicle with known aerodynamic drag characteristics. Propagating the space vehicle's state vector to an epoch time t can be described in functional form as follows:

$$\boldsymbol{x}(t) = \boldsymbol{x}(x_i(t_o), c_k) = \boldsymbol{x}(p_j), \text{ where } i = 1, \ldots, 6 \text{ and } k = 1, \ldots, m \tag{4.24}$$

where the $p_j = x_i(t_o), c_k$ is a column vector of the j solved-for parameters consisting of $x_i(t_o)$, the six-vector initial state and the c_k, m number of additional solved-for parameters. In this application the additional parameter list can be made to include the $C_{n,0}$, $S_{n,m}$, and $C_{n,m}$, the gravity harmonic coefficients. Here $j = 1, \ldots, (6+m)$. A variation of Eq. (4.24) becomes

$$\delta x_i(t) = \left[\frac{\partial x_i}{\partial p_\alpha}\right]_t \delta p^\alpha = [\mathcal{P}]_t \delta p \tag{4.25}$$

where the $[\mathcal{P}]_t$ matrix is a $6 \times (6+m)$ matrix of partial derivates evaluated for time t. Estimates for the inertial state vectors, $\hat{x}_i(t_k)$, can be made using the following coordinate transformations for each of the measured tracking data points at times $t_1, t_2, \ldots, t_n$. Here, n will be the number of tracking data measurements. Usually there will be hundreds of tracking data points available, which can provide a valid statistical basis for the solution for the solved-for parameters.

For this simplified application example, we assume that the tracking antennas can provide a very accurate ephemeris of data measurements consisting of the azimuth, $\hat{A}_Z(t_k)$, elevation, $\hat{E}_L(t_k)$, and range, $\hat{\hat{R}}(t_k)$, each taken at time t_k. These measurements transform into the topodetic position vector to the space vehicle as seen from the location of the tracking antenna as follows:

$$
\begin{aligned}
\hat{\hat{x}}_1(t_k) &= \hat{\hat{R}}(t_k)\cos\hat{E}_L(t_k)\cos\hat{A}_Z(t_k) \\
\hat{\hat{x}}_2(t_k) &= \hat{\hat{R}}(t_k)\cos\hat{E}_L(t_k)\sin\hat{A}_Z(t_k) \\
\hat{\hat{x}}_3(t_k) &= -\hat{\hat{R}}(t_k)\sin\hat{E}_L(t_k)
\end{aligned}
\tag{4.26}
$$

Using Eq. (1.14), we have the estimates for the space vehicle's position ephemeris in the ECI coordinate frame:

$$
\hat{x}_i(t_k) = a_i^{\alpha}(t_k)\hat{\hat{x}}_{\alpha}(t_k) + X_i(t_k) \tag{4.27}
$$

where $a_i^{\alpha}(t_k)$ is given by Eq. (2.111) as a function of θ, which in turn is a function of time as shown in Eq. (2.100), and the longitude and geodetic latitude of the tracking antenna. The inertial (ECI) location of the tracking antenna, $X_i(t_k)$, is given by Eq. (2.107), which is also a function of time, the longitude, the geocentric, and geodetic latitudes and H, the height above mean sea level of the tracking antenna.

A large column vector of length $6n$ is formed for the left-hand side of Eq. (4.25), which is then constructed into the matrix form as follows:

$$
\begin{pmatrix} x_i(t_1) - \hat{x}_i(t_1) \\ x_i(t_2) - \hat{x}_i(t_2) \\ \vdots \\ x_i(t_n) - \hat{x}_i(t_n) \end{pmatrix}
=
\begin{pmatrix} [\mathcal{P}]_{t_1} \\ [\mathcal{P}]_{t_2} \\ \vdots \\ [\mathcal{P}]_{t_n} \end{pmatrix}
\begin{pmatrix} \delta p_1 \\ \delta p_2 \\ \vdots \\ \delta p_{(6+m)} \end{pmatrix}
\tag{4.28}
$$

Continuing with the understanding of the matrix structures of this equation, we can simplify the matrix notation with the following representation:

$$
\delta \boldsymbol{x} = (\mathcal{P})\delta \boldsymbol{p} \tag{4.29}
$$

In this example, although the propagtion methods employed in Eq. (4.24) are very accurate when solving the space vehicle's equations of motion, topodetic velocity estimates are not being computed from tracking data measurements. Therefore, in this simple example, we select not to use any data measurements to estimate the velocity states. However, in most orbit determination processors all antenna data measurements can be used in the solution process. We can eliminate the inertial velocity variations from the analysis with the following simple square *weighting matrix*:

$$(\mathcal{W}) = \begin{pmatrix} 1 & 0 & 0 & 0 & 0 & 0 & 0 & \dots & 0 \\ 0 & 1 & 0 & 0 & 0 & 0 & 0 & \dots & 0 \\ 0 & 0 & 1 & 0 & 0 & 0 & 0 & \dots & 0 \\ 0 & 0 & 0 & 0 & 0 & 0 & 0 & \dots & 0 \\ 0 & 0 & 0 & 0 & 0 & 0 & 0 & \dots & 0 \\ 0 & 0 & 0 & 0 & 0 & 0 & 0 & \dots & 0 \\ 0 & 0 & 0 & 0 & 0 & 0 & 1 & \dots & 0 \\ \vdots & \vdots & \vdots & \vdots & \vdots & \vdots & \vdots & \ddots & \vdots \\ 0 & 0 & 0 & 0 & 0 & 0 & 0 & \dots & 0 \end{pmatrix} \tag{4.30}$$

This simple weighting matrix is formed from a $6n \times 6n$ null matrix by placing ones down the diagonal corresponding to the locations of the inertial position variations. The weighting matrix can play other roles in the statistical analysis, for instance, where each element can represent estimators for the expected relative accuracies of the individual measurements. By multiplying Eq. (4.29) by the weighting matrix, we have

$$(\mathcal{W})\delta\boldsymbol{x} = (\mathcal{W})(\mathcal{P})\delta\boldsymbol{p} \tag{4.31}$$

Solving this equation for the variations in the solved-for parameters, we have the least-squares solution:

$$[\delta\boldsymbol{p}]_N = [(\mathcal{P})^T(\mathcal{W})(\mathcal{P})]^{-1}(\mathcal{P})^T(\mathcal{W})\delta\boldsymbol{x} \tag{4.32}$$

Because this least-squares process is an overdetermined process, i.e., there are many more tracking data measurements in number than the number of solved-for parameters, the solution process is iterative. For this reason, an N subscript is attached to the variations in the solved-for parameters indicating the Nth iteration. The new estimates for the $(N + 1)$th iteration of the solved-for parameters becomes simply

$$[\boldsymbol{p}]_{N+1} = [\boldsymbol{p}]_N + [\delta\boldsymbol{p}]_N \tag{4.33}$$

The least-squares solution process is repeated until the squares of variations, δx, the *miss distances*, become a minimum, at which time the integrated trajectory is as close as possible to the tracking data measurements.

Many nonspherical gravity models for the Earth have been determined. An excellent history and discussion of the development of Earth gravity models is presented by Vallado (Ref. 8, Sec. 8.8). An example of the results from these solutions for the Earth is given in Ref. 52 (pp. 4, 5). These are the values for an $n = 4, m = 4$ gravity model for Eqs. (4.21). This model is used in many engineering flight simulation programs and is reproduced here in Table 4.1.

Using tracking data from space vehicles in lunar orbit preliminary values for the lunar gravitational field coefficients, as given in Ref. 5 (p. K7), are as shown in Table 4.2.

Table 4.1 Gravity harmonic coefficients for the Earth

Zonal coefficients	Tesseral coefficients	Sectorial coefficients
$C_{2,0} = -0.10826271E\text{-}02$	$C_{2,1} = -0.27635957E\text{-}09$	$C_{2,2} = +0.15711423E\text{-}05$
$C_{3,0} = +0.25358868E\text{-}05$	$S_{2,1} = -0.52357454E\text{-}08$	$S_{2,2} = -0.90231337E\text{-}06$
$C_{4,0} = +0.16246180E\text{-}05$	$C_{3,1} = +0.21907694E\text{-}05$	$C_{3,3} = +0.97966803E\text{-}07$
	$S_{3,1} = +0.27267074E\text{-}06$	$S_{3,3} = +0.19681077E\text{-}06$
	$C_{3,2} = +0.30466825E\text{-}06$	$C_{4,4} = -0.41542493E\text{-}08$
	$S_{3,2} = -0.21259298E\text{-}06$	$S_{4,4} = +0.63163541E\text{-}08$
	$C_{4,1} = -0.50552749E\text{-}06$	
	$S_{4,1} = -0.44125015E\text{-}06$	
	$C_{4,2} = +0.78842515E\text{-}07$	
	$S_{4,2} = +0.14818958E\text{-}06$	
	$C_{4,3} = +0.59073749E\text{-}07$	
	$S_{4,3} = -0.12140873E\text{-}07$	

Table 4.2 Preliminary values for lunar gravitational field coefficients

Zonal coefficients	Tesseral coefficients	Sectorial coefficients
$C_{2,0} = -0.2027E\text{-}03$	$C_{3,1} = +0.308E\text{-}04$	$C_{2,2} = +0.225E\text{-}04$
$C_{3,0} = -0.6E\text{-}05$	$S_{3,1} = +0.4E\text{-}05$	$C_{3,3} = +0.18E\text{-}05$
	$C_{3,2} = +0.49E\text{-}05$	$S_{3,3} = -0.3E\text{-}06$
	$S_{3,2} = +0.17E\text{-}05$	

4.2 Forces and Moments on the Flight Vehicle

4.2.1 Atmospheric Modeling

The atmospheric properties, such as fluid density, temperature, and pressure, are necessary parameters for estimating the aerodynamic forces and moments acting on aerospace vehicles during flight. If the atmosphere is assumed to be a homogeneous mix of gases with a constant molecular weight M (28.9522 kg/kmol for the Earth's atmosphere to an altitude of about 86 km), we can write the perfect gas law:

$$P = \frac{\rho R' T}{M} \tag{4.34}$$

where ρ is the fluid density, R' is the universal gas constant equal to 8.31432E+03 (N.m)/(kmol-K), and T is the gas temperature in K. The Kelvin thermometric scale is measured from absolute zero, 0 K or -273.15°C, where all molecular motion ceases. However, 1 K equals 1°C, and therefore we can write the defining relationship: $T\text{K} = T°\text{C} + 273.15°\text{C}$. Based on these assumptions, the fluid pressure as a function of altitude in the Earth's atmosphere below a height of 86 km (282,152 ft) can be approximated using the hydrostatic equilibrium relationship Ref. 53, p. 6:

$$\mathrm{d}P = -g\rho\,\mathrm{d}z \tag{4.35}$$

where dz is the differential altitude, and g is the altitude-dependent acceleration of gravity. A parcel of air in the atmosphere is being accelerated as defined by Eq. (2.109), which also defines the plumb-bob vertical line. Differentiating the perfect gas law given by Eq. (4.34) with respect to height in the atmosphere and using the hydrostatic differential equation, we have

$$\left(\frac{R'}{M}\right)\left[T\frac{\mathrm{d}\rho}{\mathrm{d}z}+\rho\frac{\mathrm{d}T}{\mathrm{d}z}\right]=-g\rho \tag{4.36}$$

This equation provides the fundamental relationship for both temperature and density as related to altitude in the atmosphere. Based on actual temperature data for the Earth's atmosphere, as shown in the *U.S. Standard Atmosphere, 1976* (Ref. 53, pp. 10, 11), the temperature can be approximated by linear relationships:

$$T = T_b + \mathcal{L}_k(z - z_b) \tag{4.37}$$

over k layers as a function of the altitude $(z - z_b)$ above the base of each layer. The k layers in the atmosphere can be defined at 0, 11, 20, 32, 47, 52, 61, 69, 79, 90, . . . to 200 km altitude. Again, however, above 90 km the temperature variations become a function of the dynamic conditions in the atmosphere.

Over these relatively small altitude variations, using similar base altitudes for the k atmospheric layers, as in Eq. (4.37), the acceleration of gravity can be closely approximated by

$$g = g_b + \mathcal{G}_k(z - z_b) \tag{4.38}$$

With these assumptions, the integral solution for density as a function of altitude, from Eq. (4.36), can be approximated by

$$\rho = \rho_{k_o} e^{-f_k(z - z_{k_b})} \tag{4.39}$$

The exponent of e can be represented by the power series

$$f_k(z - z_{k_b}) = c_{k_\alpha}(z - z_{k_b})^\alpha, \quad \text{where } \alpha = 1, m \tag{4.40}$$

The constants c_{k_α} can be determined with least-squares analysis to fit the *U.S. Standard Atmosphere* density data. Using Eqs. (4.37) and (4.39) and the perfect gas law of Eq. (4.34), the atmospheric parameters can be provided for engineering analysis flight simulations over a wide range of operational altitudes.

The speed of sound (Ref. 53, p. 18) is given by

$$V_s = \sqrt{\left(\frac{\gamma R'}{M}\right)T} \tag{4.41}$$

where $\gamma = 1.40$, the ratio of specific heat of the ambient air at constant pressure to that at a constant volume. The speed of sound is a necessary parameter to compute the freestream Mach number, $M = V/V_s$, to compute the aerodynamic forces and moments on the flight vehicle. For example, the speed of sound in ft/s becomes

$$V_s = 65.769556\sqrt{T} \tag{4.42}$$

where T is temperature in K.

Hence, for flight simulations below 86 km, the aerospace engineer can estimate the Earth's atmospheric properties using only the altitude (height) above the Earth's surface as the input parameter. However, above 86 km the Earth's atmospheric properties can no longer be accurately estimated with Eq. (4.34) due to the thermodynamics and diffusion of the atmospheric fluids (Ref. 53, p. 6). Jacchia[54] developed dynamic models for the Earth's *thermosphere*, above 90 km where the kinetic temperature begins to increase, and the *exosphere* Ref. 76, above 400 km where the temperature no longer has meaning and interplanetary space begins. Also, at these altitudes sound waves no longer exist, and Eq. (4.40) is no longer valid. The molecular weight of the constituent gases, such as nitrogen and oxygen, varies and must be summed to estimate the total atmospheric molecular weight. The atmospheric density becomes a function of the following conditions and their variations (Ref. 54, p. 17): 1) the solar cycle, 2) daily changes in solar activity on the visible disk of the sun, 3) daily or diurnal variations, 4) geomagnetic activity, 5) seasonal-latitude variations, 6) semi-annual variations, and 7) rapid density fluctuations probably connected with gravity waves.

When using the Jacchia atmospheric models, the following additional input parameters are required to estimate the atmospheric density, which will include the effects due to the preceding variations:

1) The *Julian date of the epoch* of the input state vector.

2) The *inertial position vector to the sun* at the epoch.

3) The *Earth-centered, Earth-fixed (ECF) state vector* or the longitude, latitude, and height above the Earth's surface of the space vehicle.

4) The F10.7 cm wavelength radio wave intensity, which is called the *solar flux*, at a time as close as possible to the epoch of the input state vector. The values for the F10.7 solar flux and the Earth's geomagnetic index are distributed daily by the National Oceanic and Atmospheric Administration (NOAA) from the National Geophysical Data Center located in Boulder, Colorado. The solar flux (F10.7) is given in units of $10^{-22}\ \mathrm{W/(m^2\ Hz)}$. The unit Hz was mentioned earlier in the text, where one Hz is equal to one cycle per second.

5) The *geomagnetic index*, given likewise at the epoch time. The geomagnetic index (amplitude) a_p is given in units of 10^{-9} (kg s)/m or 10^{-9} T, and is called one gamma.

A mathematical discussion of the development of the Jacchia models is beyond the scope of this text. However, an excellent presentation of this analysis is given in Ref. 8 (Appendix B). Many of these computer algorithms are available to the aerospace engineer and are necessary for accurate flight simulations to study the motion of space vehicles.

4.2.2 Aeronautical Nomenclature

In a similar manner as the four-DOF motion introduced in Sec. 2.3.3, we now extend the analysis to describe motion of the airframe with six-degrees of freedom, i.e., free to translate in three directions and free to rotate on the three coordinate axes. Because of the number of parameters required to describe flight vehicle six-DOF motion, it is necessary to define the database and the engineering symbolism that is generally used. The definitions of parameters presented in this section are general in nature and are applied to any airframe type. However, the analysis of the

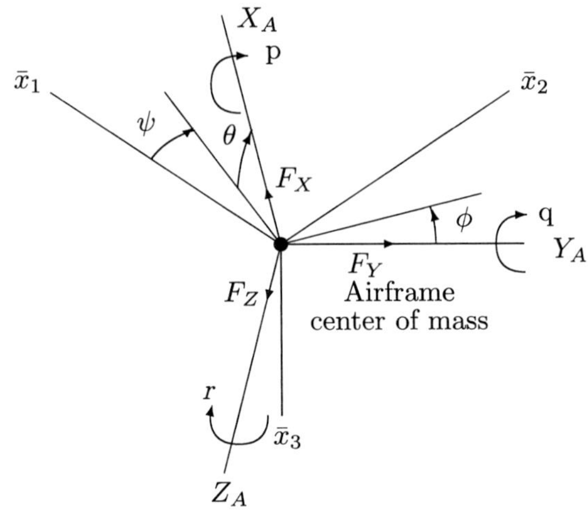

Fig. 4.2 Aeronautical engineering angles and the body axis rotation rates.

motion of airplanes, ascent rocket-powered vehicles, entry vehicles, etc., involves different airframe structures, and hence the predominant aerodynamic parameters will differ.

Note that these general discussions, especially concerning six-DOF motion simulation, are intended for computer database application. Using modern computer systems, the level of detail that can be included far exceeds the analysis capabilities of hand calculations, where many approximations would have to be made to study the motion of the airframe. Figure 4.2 helps to define the forces and moments acting on the airframe using standard aeronautical engineering nomenclature:

To be consistent with the generalized notation used in the text, we will continue to show its correspondence with *standard aeronautical engineering notation*. The standard engineering notation for body axis forces and moments and their coefficients (Ref. 19, p. 176) are shown in Table 4.3.

In the text the unbarred frames are generally the inertial frames, i.e., the x axes. The barred frames, the $\bar{x}$ frames, are the attached frames or non-inertial frames, as in this case, the airframe coordinates or a ground-based coordinate frame. Lastly, the double-barred frames, the $\bar{\bar{x}}$ frames, can represent the coordinate frames relative to the barred frames. With these definitions in mind, the vector definitions shown in Table 4.4 will help to characterize the corresponding relationships in the notations.

Table 4.3 Standard engineering notation for body axis forces and moments and their coefficients

Airframe axis	Forces		Moments		
X	F_X	C_X	Roll axis,	$\mathcal{L}$	C_l
Y	F_Y	C_Y	Pitch axis,	$\mathcal{M}$	C_m
Z	F_Z	C_Z	Yaw axis,	$\mathcal{N}$	C_n

Table 4.4 Corresponding aeronautical parameters

Airframe	Aeronautical notation	Text equivalent
Axes	(X_A,Y_A,Z_A)	$(\bar{x}_1,\bar{x}_2,\bar{x}_3)$
Local velocity	(u, v, w)	$(\bar{\bar{v}}_1, \bar{\bar{v}}_2, \bar{\bar{v}}_3)$
Euler angles	(ψ,θ,ϕ)	$(\theta_1,\theta_2,\theta_3)$
Axis rotation rates	(p, q, r)	$(\bar{\omega}_1,\bar{\omega}_2,\bar{\omega}_3)$
Axis forces	(F_X,F_Y,F_Z)	$(\bar{\bar{f}}_1, \bar{\bar{f}}_2, \bar{\bar{f}}_3)$
Axis moments	$(\mathcal{L}, \mathcal{M}, \mathcal{N})$	$(\bar{\bar{n}}_1, \bar{\bar{n}}_2, \bar{\bar{n}}_3)$

The yaw-pitch-roll Euler angle set (ψ, θ, ϕ) defines the airframe attitude orientation with the reference ground-based coordinates. The attitude transformation matrix is given in Appendix A.10. Positive rotations and moments are generally defined using the right-hand rule. For example, a positive roll rate p, about the $\bar{\bar{x}}_1$ axis, is a roll to the right; a positive pitch rate q, about the $\bar{\bar{x}}_2$ axis, is a pitch up; and a positive yaw rate r, about the airframe $\bar{\bar{x}}_3$ axis, is a yaw to the right.

The airplane force coefficients along each airframe axis are transformed from the *wind axis* to the body axis via the *airplane angle of attack* as follows:

$$
\begin{aligned}
C_X &= C_L \sin\alpha - C_D \cos\alpha \\
C_Y &= C_Y \\
C_Z &= -C_L \cos\alpha - C_D \sin\alpha
\end{aligned}
\tag{4.43}
$$

Notice the sign convention here; a positive angle of attack gives a negative lift component, i.e., in the negative $\bar{\bar{x}}_3$ direction. In the analysis discussions of Sec. 2.3.3 and here in Eq. (4.43), the airplane lift and drag coefficients (representing those forces perpendicular and parallel to the relative velocity vector, respectively) must include the lift and drag due to the wing as well as that generated by the body and tail surfaces.

In summary, the total body axis forces [similar to the forces given in Eq. (2.127)] and total body axis moments acting on the airframe are generally estimated with the following:

$$
\begin{aligned}
\bar{\bar{f}}_1 &= F_X = \bar{\bar{T}}_X + C_X\bar{q}S \\
\bar{\bar{f}}_2 &= F_Y = \bar{\bar{T}}_Y + C_Y\bar{q}S \\
\bar{\bar{f}}_3 &= F_Z = \bar{\bar{T}}_Z + C_Z\bar{q}S \\
\bar{\bar{n}}_1 &= \mathcal{L} = \mathcal{L}_T + C_l\bar{q}Sb \\
\bar{\bar{n}}_2 &= \mathcal{M} = \mathcal{M}_T + C_m\bar{q}S\bar{c} \\
\bar{\bar{n}}_3 &= \mathcal{N} = \mathcal{N}_T + C_n\bar{q}Sb
\end{aligned}
\tag{4.44}
$$

where the $(\bar{\bar{T}}_X,\bar{\bar{T}}_Y,\bar{\bar{T}}_Z)$ is the engine thrust force vector components and the $(\mathcal{L}_T,\mathcal{M}_T,\mathcal{N}_T)$ represents the torque vector components due to engine thrust.

Engine thrust modeling will be discussed in the next sections. The preceding airframe parameters are defined as follows:

$\bar{q} = (1/2)\rho\bar{v}^2$, the dynamic pressure of the air relative to the airframe
S = the reference wing area
b = the reference wing span
$\bar{c}$ = the mean chord length of the wing

The body axis forces in Eqs. (4.44) are transformed into the Earth-fixed coordinates and become the driving function in Eq. (2.124) for the solution of the translational dynamics of the airframe. Likewise, the body axis moments in Eqs. (4.44) become the driving torques in Euler's equation, given by Eq. (3.77), for the solution of the rotational dynamics of the airframe. The simultaneous solution of these six differential equations provides an estimate for the motion of the airframe.

The non-dimensional coefficients in Eqs. (4.44) are functions of the dynamic flight parameters, the control parameters, and the physical dimensions and locations of the aerodynamic surfaces. The coefficients are specific for each aerospace vehicle, and they are the results of lengthy engineering studies involving wind-tunnel testing, flight testing, and in some cases, computer finite element analysis. However, once the aerodynamic force and moment coefficients have been determined, actual flight characteristics can be simulated, and flight performance can be estimated over a large part of the aircraft's flight envelope.

4.2.3 Fluid Dynamic Forces and Moments from Finite Element Modeling

The analysis presented in this section is intended to be conceptual in nature, since a detailed discussion of the computational methods of finite element aerodynamics is beyond the scope of this book. However, it is very important to recognize this new technolgy and the aeroengineer's ability to predict the forces and moments acting on an arbitrary airframe using finite element numerical methods. These analyses can closely estimate the airplane's flight characteristics and stability. Studies can be done in the airframe design phase, long before the flight vehicle is built and test flown. Potential flow analysis and finite element solution methods have become possible along with the advances in large-scale mainframe digital computing systems. Prior to 1965, the application of these solution methods was simply not possible. New finite element aerodynamic modeling techniques are still under development today.

The sum of all fluid dynamic pressure forces acting on the surface areas of the flight vehicle results in the forces and moments on the airframe that affect its translational and rotational motion. We can represent the three-dimensional surface of the airframe by dividing its surface area into small panels or finite area elements. Each of the panels are tangent to the vehicle's surface at selected center-of-pressure points over the entire surface of the flight vehicle. Figure 4.3 depicts an arbitrary flight vehicle showing a simplified geometric panel representation and where the pressure forces act on the ϱth panel. The variation of the symbol ρ is used to identify each finite area panel so as not to conflict with the standard aeronautical use of the Greek symbols or to be confused with vector component notations.

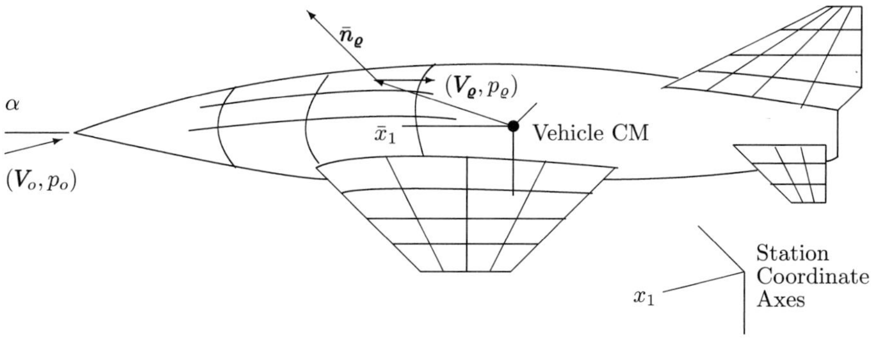

Fig. 4.3 Finite element paneling of an arbitrary flight vehicle.

The force on the ϱth panel located at $\bar{\boldsymbol{x}}_\varrho$ relative to the center of mass of the airframe simply becomes

$$\bar{\boldsymbol{f}}_\varrho = (p_o - p_\varrho)A_i\bar{\boldsymbol{n}}_\varrho \tag{4.45}$$

where p_o is the ambient pressure, and p_ϱ is the pressure on the outside of the panel or on the airframe surface at the point $\bar{\boldsymbol{x}}_\varrho$. The vector $\bar{\boldsymbol{n}}_\varrho$ is the unit normal to the airframe surface, again at the point $\bar{\boldsymbol{x}}_\varrho$.

For a constant geo-potential height and temperature, the total energy in a fluid stream flow is a function of its instantaneous kinetic and pressure energies. Hence, the total energy from one point, (V_o, p_o), to another point, (V_ϱ, p_ϱ), is constant and is given by equating the sums of its pressure energy and kinetic energy as follows:

$$p_o + (1/2)\rho V_o{}^2 = p_\varrho + (1/2)\rho V_\varrho{}^2 \tag{4.46}$$

This total energy relationship in *subsonic/incompressible flow* forms *Bernoulli's equation*, and from this we have the definition of the pressure coefficient:

$$C_{p_\varrho} = \frac{(p_\varrho - p_o)}{\bar{q}} = 1 - \left(\frac{V_\varrho}{V_o}\right)^2 \tag{4.47}$$

The pressure coefficient is used throughout aerodynamics to describe the relative pressures on airframe surfaces. Generally, a negative pressure coefficient causes an outward force on the panel. Using the pressure coefficient in Eq. (4.47), we can approximate the force on the ϱth finite element panel by

$$\bar{\boldsymbol{f}}_\varrho = -C_{p_\varrho}\bar{q}A_\varrho\bar{\boldsymbol{n}}_\varrho \tag{4.48}$$

Notice from Eq. (4.47) that the pressure coefficient on the ϱth panel can be estimated by determining the velocity magnitude of the fluid passing over the panel. Using large-scale digital computer systems and numerical methods, the engineer can approximate a vector field, $\boldsymbol{V}_\varrho = \boldsymbol{V}_\varrho(\bar{\boldsymbol{x}}_\varrho, t)$, for each panel element located at

$\bar{\boldsymbol{x}}_{\varrho}$ in the field. For example, from Fig. 4.3 we can write the air/fluid velocity over the ϱth panel in station coordinates:

$$\boldsymbol{V}_{\varrho} = \boldsymbol{v}_{\varrho} + \Delta \boldsymbol{V}_{\varrho} \tag{4.49}$$

where $\boldsymbol{v}_{\varrho}$, is the fluid velocity over the ϱth panel due to the geometry of the location of the panel itself and $\Delta \boldsymbol{V}_{\varrho}$, is the ϱth perturbation in the fluid motion caused by the aerodynamic flow field around the entire airframe.

The station coordinates for any point on the vehicle are given by $x_i = a_i^{\alpha} \bar{x}_{\alpha} + X_i$ (here the a matrix is the body-axis-to-station-coordinate transformation matrix), and the ϱth panel can be located in body/vehicle coordinates by $\bar{x}_i = b_i^{\alpha} \bar{\bar{x}}_{\alpha} + \bar{X}_i$, where the b matrix is the ϱth panel orientation with respect to vehicle coordinates, and $\bar{X}_i$ is its position in body axis coordinates. The $\bar{\bar{x}}_{\alpha}$ are the position components of any point in the panel coordinate frame. Hence, any position on the ϱth panel in station coordinates is given by $x_i = a_i^{\alpha}(b_{\alpha}^{\beta} \bar{\bar{x}}_{\beta} + \bar{X}_{\alpha}) + X_i$. Differentiating this equation, remembering that the panel is fixed in body axis coordinates and that we are interested only in the defined center-of-pressure point, i.e., where $\bar{\bar{x}}_{\beta} = 0$ on the panel, we have the fluid velocity at the ϱth panel location, $v_i = a_i^{\alpha} \bar{W}_{\alpha}^{\beta} \bar{X}_{\beta} + V_{o_i}$. From the body axis rotation rates, as shown in Table 4.4, we have the body-axis-rotation-rate matrix:

$$(\bar{W}) = \begin{pmatrix} 0 & -r & q \\ r & 0 & -p \\ -q & p & 0 \end{pmatrix} \tag{4.50}$$

Hence, we realize that the vector $\boldsymbol{v}_{\varrho}$ in Eq. (4.49) must be the sum of the fluid velocity due to body axis rotations relative to the station coordinates and the speed of the vehicle in station coordinates, $\boldsymbol{V}_o$.

From this analysis, Eq. (4.49) in component form becomes

$$V_{\varrho i} = \left(a_i^{\alpha} \bar{W}_{\alpha}^{\beta} \bar{X}_{\beta} + V_{o_i}\right) + \Delta V_i \tag{4.51}$$

From the preceding discussion, notice that the matrix $(c)^T = (b)^T (a)^T$ will transform vectors from the station coordinate frame into the local ϱth panel frame. Hence, by multiplying Eq. (4.51) by the $(c)^T$ matrix, we have an estimating relationship for the local fluid velocity over the ϱth panel:

$$\begin{pmatrix} u_{\varrho} \\ v_{\varrho} \\ 0 \end{pmatrix} = \begin{pmatrix} U_{\varrho} \\ V_{\varrho} \\ W_{\varrho} \end{pmatrix} + \begin{pmatrix} \Delta u_{\varrho} \\ \Delta v_{\varrho} \\ \Delta w_{\varrho} \end{pmatrix} \tag{4.52}$$

The uppercase $(U_{\varrho}, V_{\varrho}, W_{\varrho})$ here simply represents the known geometric fluid flow components on the ϱth panel, i.e., from the terms in the parentheses of Eq. (4.51). Notice that the z component of velocity over the panel is constrained to $w_{\varrho} = 0$, since there can be no fluid flow through the surface of the panel. This gives the three velocity component equations:

$$\begin{aligned} u_{\varrho} &= U_{\varrho} + \Delta u_{\varrho} \\ v_{\varrho} &= V_{\varrho} + \Delta v_{\varrho} \\ \Delta w_{\varrho} &= -W_{\varrho} \end{aligned} \tag{4.53}$$

The Δu_ϱ and Δv_ϱ terms are the variations in the component velocities that are the sums of the aerodynamic influences of all of the other panels of the potential flowfield representing the flight vehicle. Thus, we have $V_\varrho{}^2 = u_\varrho{}^2 + v_\varrho{}^2$ and hence an estimation for the pressure coefficient on the ϱth panel from Eq. (4.47). Note that if the ϱth panel geometry is defined with its sides nearly parallel to the estimated locate fluid velocity flow; Δv_ϱ is very small and Δu_ϱ becomes the predominant solution parameter for V_ϱ, namely $V_\varrho \approx u_\varrho$.

Potential flow models using a matrix of *influence coefficients*[55] can be made to converge on the airframe's vector velocity field. The effects of one panel on another in the flowfield are called the *aerodynamic influence coefficients*.

The velocity field can also be estimated from the first integral solution of the equations of motion for the finite fluid elements. Starting with Eq. (2.114), the equations of motion for the small fluid elements can be developed. If we express the finite element mass as $\mathrm{d}m = \rho \mathrm{d}\nu$, where ν is the volume of the fluid element enclosed by some surface areas, then Eq. (2.114) in vector form can be written

$$\rho\,\mathrm{d}\nu \left(\frac{\mathrm{d}\boldsymbol{V}_\varrho}{\mathrm{d}t}\right) = \mathrm{d}\boldsymbol{F}_{surf} + \rho\,\mathrm{d}\nu \begin{pmatrix} 0 \\ 0 \\ \bar{g}_s \end{pmatrix} \tag{4.54}$$

where $\mathrm{d}\boldsymbol{F}_{surf}$ is the differential pressure forces acting on the finite element surface areas. When the frictional sheer forces between layers of air moving at different speeds (viscous flow effects), the conservation of momentum and the forces due to rotations in the air (vorticity effects) are included in $\mathrm{d}\boldsymbol{F}_{surf}$ terms, Eq. (4.54) becomes the *Navier-Stokes equations* (Ref. 56, Sec. 15.4). The numerical solutions[57] to the Navier-Stokes equations, as applied to airplane finite aerodynamics, is an important technology that is still being developed today.

For subsonic flight, as discussed earlier, the pressure coefficient C_{p_ϱ} at the ϱth panel is influenced by the airflow over all of the other panels on the surface of the airframe. Likewise, all of the other panels are affected by the airflow over the ϱth panel. From the dynamics of fluids (Ref. 56, Sec. 9.1), we recall that *pressure effects in a fluid are transmitted at the speed of sound.* In supersonic flight *shock waves* are formed where the abrupt changes in fluid flow direction occur due to the motion of the oncoming flight vehicle.

Assume for the moment that we have only three panel areas, i,j, and k. In supersonic flight the airflow over the ith panel can only affect those panels behind the ith panel's Mach cone, as shown in Fig. 4.4. For example, from Fig. 4.4, the ith panel *does not influence* the velocity field of the jth panel; however, the ith panel *does influence* the velocity flow over the kth panel.

Finite element computational methods, as well as other aerodynamic analysis methods, use different analyses depending on the flight Mach number regions being studied. As we have shown from the preceding discussion, the local Mach number at different points on the surface of the flight vehicle will vary considerably from the actual flight Mach number. Because the precise definition of these regions depends on the local surface Mach numbers, the following flight regimes will vary based on the shape and geometry of the flight vehicle itself.

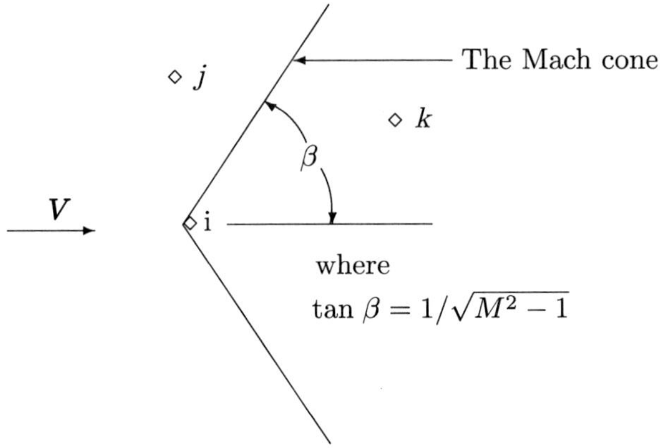

Fig. 4.4 Panel in front of and behind the Mach cone.

These regions can be approximately defined by the flight Mach number range as follows:

1) *Subsonic*, $M < 0.80$; all aerodynamic surfaces influence all other flight vehicle surfaces, and all surface Mach numbers are less than one, i.e., no shock waves are formed.

2) *Transonic*, $0.80 < M > 1.20$; shock waves are formed as the fluid accelerates to Mach 1 over some parts of the flight vehicle surfaces.

3) *Supersonic*, $M > 1.20$; all local surface fluid flows are supersonic; shock waves are formed and emanate from all leading surface areas of the airframe by the angle β.

4) *Hypersonic*, $M \gg 5$; the leading shock waves are forced down into the fluid flow boundary layer (the boundary of the lower speed airflow immediately adjacent to the airframe surfaces).

5) *Newtonian*, $M > 8$?; the shock waves are now on all of the airframe surfaces exposed to the oncoming slip stream.

6) *Molecular*; at orbital speeds where the Mach number is ill-defined and fluid molecules or ions behave like small pool balls when surface contact is made.

To briefly summarize the preceding discussions, 1) the surface of a flight vehicle can be geometrically represented by n_ϱ finite area panels, each parallel to the actual flight vehicle's contoured shape; 2) the velocity field solution estimates the changes to the vehicle's flight velocity yielding the fluid's relative velocity at each panel's center of pressure, $\boldsymbol{V}_\varrho$, which is constrained to be parallel to each of finite area panels; 3) the pressure coefficients C_{p_ϱ} are then estimated at the defined center-of-pressure points on the finite element area panels; and 4) the pressure force on each panel is estimated. These forces are summed, and the body axis force coefficients for the flight vehicle become

$$(\bar{q}S)\begin{pmatrix} C_X \\ C_Y \\ C_Z \end{pmatrix} = \sum_{\varrho=1}^{n_\varrho} \bar{f}_\varrho \tag{4.55}$$

The airplane lift and drag cofficients can now be determined using the body-axis-to-wind-axis transformation matrix given by Eqs. (2.122), and can be estimated as follows:

$$\begin{pmatrix} C_D \\ C_Y \\ C_L \end{pmatrix} = \begin{pmatrix} \cos\alpha\cos\beta & \cos\alpha\sin\beta & \sin\alpha \\ -\sin\beta & \cos\beta & 0 \\ \sin\alpha\cos\beta & -\sin\alpha\sin\beta & \cos\alpha \end{pmatrix} \begin{pmatrix} C_X \\ C_Y \\ C_Z \end{pmatrix} \tag{4.56}$$

The body axis moment coefficients are likewise estimated with

$$(\bar{q}S) \begin{pmatrix} bC_l \\ \bar{c}C_m \\ bC_n \end{pmatrix} = \sum_{\varrho=1}^{n_\varrho} (\bar{\boldsymbol{x}}_\varrho \times \bar{\boldsymbol{f}}_\varrho) \tag{4.57}$$

By varying the initial flight parameters, i.e., the starting boundary conditions, estimates for the aerodynamic derivatives may be found. For example, if two solutions for the pressure coefficients are performed, one with the flight vehicle angle of attack of α and the other at $\alpha = \alpha + \delta\alpha$, the flight vehicle pitching moment derivative with respect to α is estimated as follows:

$$C_{m_\alpha} = \frac{\partial C_m}{\partial \alpha} = \frac{C_m(\alpha + \delta\alpha) - C_m(\alpha)}{\delta\alpha} \tag{4.58}$$

The flight vehicle change in the aerodynamic pitching moment with respect to a change in angle of attack is an important dynamic stability parameter. For a *stable flight vehicle*, there should be a pitch down moment response to an increase in angle of attack, i.e., the value for C_{m_α} in Eq. (4.58) should be negative.

4.2.4 Subsonic Aerodynamic Lifting Surface Model

This section presents a brief outline of the methods and parameters used in describing the magnitudes of the forces and moments acting on a two-dimensional aerodynamic lifting surface in subsonic flow. This method can be used to approximate an entire wing or tail surface characteristics without using finite element paneling to describe the lifting surface. These methods were used prior to high-speed computer systems and are still used today for preliminary aircraft design and engineering analysis to estimate airplane motion. The ϱth aerodynamic lifting surface of the airplane can be represented by the force and moment diagram shown in Fig. 4.5. Here, α is the dynamic parameter, the airplane angle of attack as given in Eq. (2.120). The aerodynamic surface structural mounting angle, the *angle of incidence*, is i_ϱ, which is exaggerated for clarity. The aerodynamic surface angle of attack becomes

$$\alpha_\varrho = \alpha + i_\varrho \tag{4.59}$$

The control surface deflection angle shown in Fig. 4.5 is δ_ϱ. A positive deflection angle is normally downward, which causes an increase in the lift created by the aerosurface. The lift and drag forces, as described in Sec. 2.3.3, respectively, and now the moment magnitudes, can be defined to be acting only on the ϱth

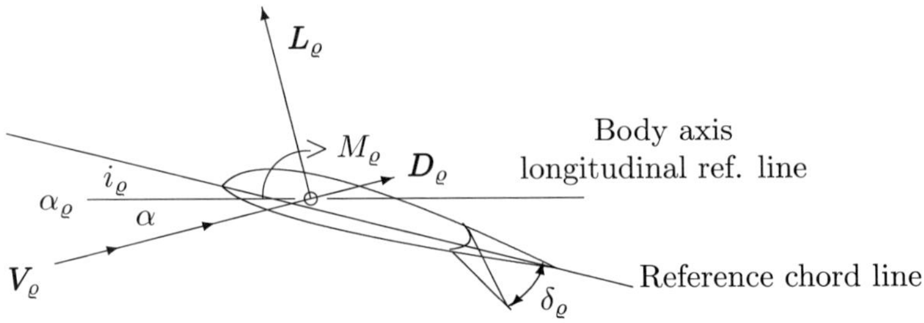

Fig. 4.5 Two dimensional aerodynamic forces and moments.

aerodynamic surface, as shown in Fig. 4.5. The magnitudes of these forces and the pitching moment, which are considered to act at the center of pressure of the lifting surface, can be estimated as follows:

$$\begin{aligned} L_\varrho &= C_{L_\varrho} \bar{q}_\varrho S_\varrho \\ D_\varrho &= C_{D_\varrho} \bar{q}_\varrho S_\varrho \\ M_\varrho &= C_{m_\varrho} \bar{q}_\varrho S_\varrho \bar{c}_\varrho \end{aligned} \tag{4.60}$$

The dynamic pressure felt at the ϱth lifting surface is $\bar{q}_\varrho = (1/2)\rho | \mathbf{V}_\varrho |^2$. If the lift on the aerodynamic surface is limited to be a function of the angle of attack and the control deflection angle only, the coefficients for each surface become

$$\begin{aligned} C_{L_\varrho} &= C_{L_{\varrho o}} + C_{L_{\alpha_\varrho}}(\alpha_\varrho - \alpha_{\varrho_o}) + C_{L_{\delta_\varrho}}(\delta_\varrho - \delta_{\varrho_o}) \\ C_{D_\varrho} &= C_{D_{\varrho o}} + \frac{C_{L_\varrho}^2}{\pi \epsilon AR} + C_{D_{\delta_\varrho}}(\delta_\varrho - \delta_{\varrho_o}) \\ C_{m_\varrho} &= C_{m_{\varrho o}} + C_{m_{\delta_\varrho}}(\delta_\varrho - \delta_{\varrho_o}) \end{aligned} \tag{4.61}$$

Note that three levels of subscripting are used here, denoting the coefficients and angle parameters specifically for the ϱth aerodynamic surface.

The aerodynamic forces from Eq. (4.60) can be transformed to body axis coordinates using the airplane angle of attack α and the wind-axis-to-body-axis transformation matrix given by Eq. (2.123), as follows:

$$\begin{aligned} F_{X_\varrho} &= +L_\varrho \sin\alpha - D_\varrho \cos\alpha \\ F_{Z_\varrho} &= -L_\varrho \cos\alpha - D_\varrho \sin\alpha \end{aligned} \tag{4.62}$$

Remember that the positive airplane z axis is defined downward.

4.2.5 Simulating Flight Vehicle Engine Thrust: The Subsonic Airplane Propeller

In this section, we will briefly outline the fundamental methods that can be used in computer simulations to model the thrust forces produced by subsonic airplane propellers. Looking down on a cross section of a conventional rotating propeller blade (turning in the direction defined by the right-hand rule), Fig. 4.6 shows a diagram of fundamental dynamic parameters and forces. The station relative flight vehicle's velocity vector along with the effects of the rotating propeller blade can be transformed into relative velocity wind components perpendicular and parallel to the motion of the propeller blades, represented by the vector $\boldsymbol{V}_{rw}$ in the figure. These components will vary as the propeller blades turn about the propeller hub and are functions of the translational and rotational motion of the airframe as well as the rotating motion of the propeller blades. The propeller blade relative fluid velocity vector will be discussed in some detail in the following paragraphs. The blade relative wind velocity vector subtends an instantaneous angle δ with the rotating propeller plane, as shown in Fig. 4.6. $\boldsymbol{V}_{\text{rot}}$ is the speed of the air due to the rotation of the propeller plus the small velocity component of the relative wind in the propeller plane and is given by

$$V_{\text{rot}} = \frac{\pi(RPM)}{30} r_{\text{sec}} + V_{rw} \sin\delta \tag{4.63}$$

RPM is the rotational speed of the propeller in rpm, and r_{sec} is the radius distance the propeller airfoil section is from the axis of rotation of the propeller. The vector sum of these two components is V_{sec}, the speed of the air passing the propeller airfoil section:

$$V_{\text{sec}} = \sqrt{V_{\text{rot}}^2 + (V_{rw}\cos\delta)^2} \tag{4.64}$$

The relative wind angle α_{sec} at the section radius r_{sec} is simply

$$\alpha_{\text{sec}} = \tan^{-1}\left(\frac{V_{rw}\cos\delta}{V_{\text{rot}}}\right) \tag{4.65}$$

The relative wind angle of attack of the airfoil section becomes $\alpha = i_{\text{sec}} - \alpha_{\text{sec}}$. Here the section angle of attack is estimated from the propeller geometry only. The actual section angle of attack is influenced by the change in air motion prior to encountering the propeller blades.

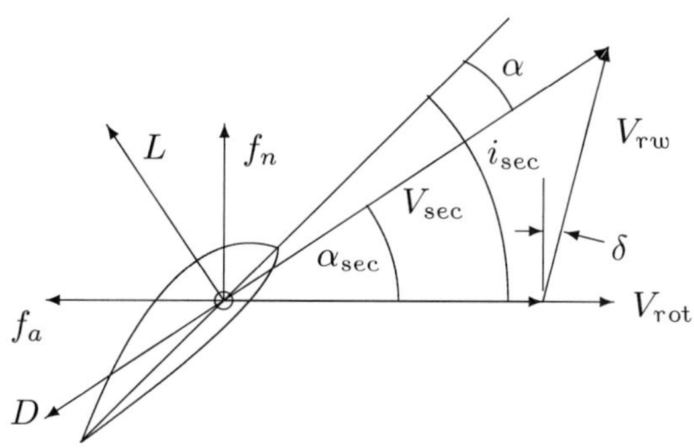

Fig. 4.6 Propeller section dynamic parameters and forces.

The propeller section angle of incidence as a function of r_{sec}, which can be used to define the required *propeller twist angles* for an airplane operational speed range, becomes

$$i(r_{\text{sec}}) = \alpha_{\text{desr}} + \tan^{-1}\left(\frac{V_{rw}\cos\delta}{V_{\text{rot}}}\right) \tag{4.66}$$

where α_{desr} is the desired propeller angle of attack. The fixed pitch propeller twist angles may be estimated using $\delta = 0$ for an optimum design. The Mach number at the propeller airfoil section is

$$M_{\text{sec}} = \frac{V_{\text{sec}}}{V_s} \tag{4.67}$$

where V_s is the local speed of sound given in Eq. (4.42).

We can simulate the sum of all propeller section forces with total lift and drag forces acting at a center-of-pressure point located at some radius $\bar{r}$ from the axis of rotation of the propeller. Looking at a typical airplane propeller blade, the total lift and drag forces (in airplane body axis coordinates) acting on the propeller can be visualized using the diagram shown in Fig. 4.7. The thrust from the propeller, which is the summation of propeller forces directed perpendicular to the rotation plane of the propeller, can be estimated using Eq. (4.62) as follows:

$$\bar{\bar{T}}_1 = \sum_{i=1}^{N}(\bar{C}_{L_i}\sin\alpha_{\text{sec}} - \bar{C}_{D_i}\cos\alpha_{\text{sec}})\bar{q}_p\bar{S} = \sum_{i=1}^{N}\bar{\bar{f}}_{n_i} \tag{4.68}$$

where N is the number of propeller blades. $\bar{C}_{L_i}$ and $\bar{C}_{D_i}$ are functions of the angle of attack of the propeller, $\alpha = i_{\text{sec}} - \alpha_{\text{sec}}$, as shown Fig. 4.6. The dynamic pressure of the rotating propeller, $\bar{q}_p$, can be estimated by $(1/2)\rho V_{\text{sec}}^2$. $\bar{S}$ is the reference area of the propeller blade. In this simulation analysis, the propeller angle of incidence i_{sec} can be considered constant when simulating a *fixed-pitch propeller*.

When the airplane engine is mounted on the airframe, the propeller shaft pointing direction, shown in Fig. 4.7, can be described by two rotation angles, θ_1 and θ_2, relative to the airplane body axes. The *propeller-plane-coordinates-to-airplane-body-axis-coordinates* transformation matrix is given by $(b) = Z(\theta_1)Y(\theta_2)$. When the thrust forces are perpendicular to the propeller plane, the thrust vector in the airplane body axis coordinates becomes

$$\bar{\boldsymbol{T}} = \begin{pmatrix} \cos\theta_1\cos\theta_2 & -\sin\theta_1 & \cos\theta_1\sin\theta_2 \\ \sin\theta_1\cos\theta_2 & \cos\theta_1 & \sin\theta_1\sin\theta_2 \\ -\sin\theta_2 & 0 & \cos\theta_2 \end{pmatrix}\begin{pmatrix} \bar{\bar{T}}_1 \\ 0 \\ 0 \end{pmatrix} = \bar{\bar{T}}_1\begin{pmatrix} \cos\theta_1\cos\theta_2 \\ \sin\theta_1\cos\theta_2 \\ -\sin\theta_2 \end{pmatrix} \tag{4.69}$$

Note, as mentioned earlier, the propeller blade relative geometric wind component, V_{rw} in Eq. (4.64), can be estimated using the following analysis: 1) the coordinates

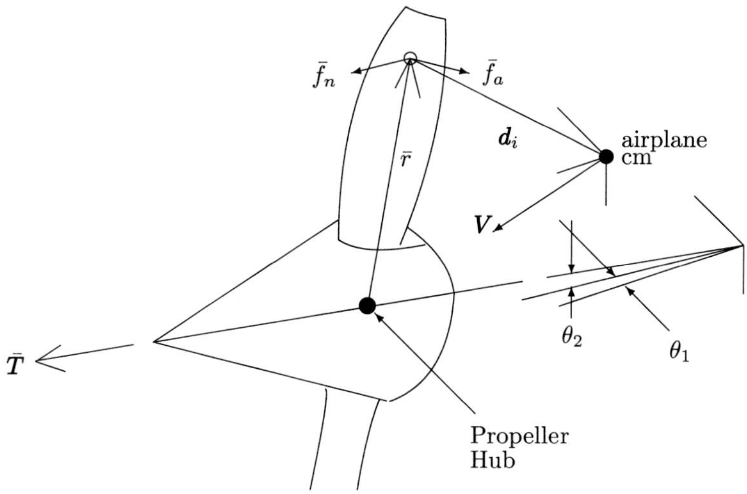

Fig. 4.7 Visualizing the forces acting on the propeller blade.

of the propeller blade center-of-pressure point in propeller plane coordinates is given by $(0, \bar{r}\cos\theta_r, \bar{r}\sin\theta_r)$, where θ_r is the instantaneous blade angle relative to the propeller plane y axis; 2) the velocity of the propeller blade center-of-pressure point in propeller plane coordinates is $(0, -\bar{r}\bar{\bar{\omega}}_r\sin\theta_r, \bar{r}\bar{\bar{\omega}}_r\cos\theta_r)$, where $\bar{\bar{\omega}}_r$ is the propeller rotation rate in rad/s; 3) using the same analysis that was used to define Eq. (4.51), we can write the station relative position components of any point in propeller blade coordinates, $x_i = a_i^\alpha b_\alpha^\beta (c_\beta^\gamma \bar{\bar{\bar{x}}}_\gamma + \bar{\bar{X}}_\gamma) + a_i^\alpha \bar{X}_\alpha + X_i$, where the c matrix is the single-axis rotation matrix, propeller-blade-coordinates-to-propeller-plane-coordinates transformation matrix; 4) differentiate this equation with respect to time and remember that we are interested only in the propeller blade center-of-pressure point, i.e., $\bar{\bar{\bar{x}}}_\gamma = 0$, and that the propeller hub is stationary relative to the flight vehicle's body axis frame; and finally 5) multiplying by $(c)^T(b)^T(a)^T$, we have the geometric fluid velocity vector in propeller blade coordinates as shown in Fig. 4.6:

$$V_{r\omega} = (c)^T \left[((b)^T(\bar{W})(b)) \begin{pmatrix} 0 \\ \bar{r}\cos\theta_r \\ \bar{r}\sin\theta_r \end{pmatrix} + \begin{pmatrix} 0 \\ -\bar{r}\bar{\bar{\omega}}_r\sin\theta_r \\ \bar{r}\bar{\bar{\omega}}_r\cos\theta_r \end{pmatrix} + ((b)^T(\bar{W}))\bar{X} + ((b)^T(a)^T)V \right] \tag{4.70}$$

The vector V in Eq. (4.70) must include the atmospheric winds as described in Eq. (2.119). The propeller blade relative fluid velocity causes the downward-going propeller blade (for positive airplane angles of attack) to have a greater angle of attack and hence experiences greater lift forces in contrast with the upward-going propeller blade. This produces a negative moment (torque) about the airplane body z axis. This torque plays an important role in determining the *critical engine* for multi-engine airplanes during engine-out flight conditions. The critical engine is defined as the failed engine that forces the use of the engine that will cause the maximum z axis torque on the airframe. The pilot must balance this torque by using rudder and aileron deflections to maintain control of the airplane during the engine-out configuration.

The instantaneous moments transferred to the airplane body axis become

$$\begin{pmatrix} \bar{N}_1 \\ \bar{N}_2 \\ \bar{N}_3 \end{pmatrix} = \begin{pmatrix} \mathcal{L} \\ \mathcal{M} \\ \mathcal{N} \end{pmatrix} = \sum_{i=1}^{N} \left(\boldsymbol{d}_i \times \bar{\boldsymbol{f}}_{n_i} + \bar{C}_{m_i} \bar{q}_p \bar{S} \bar{c}_i \left(\frac{\bar{\boldsymbol{r}}_i}{\bar{r}_i} \right) \right) - \begin{pmatrix} \bar{W}_1^{\alpha} b_{\alpha}^{\beta} \bar{\bar{I}}_{\beta}^{\gamma} \bar{\bar{\omega}}_{\gamma} \\ \bar{W}_2^{\alpha} b_{\alpha}^{\beta} \bar{\bar{I}}_{\beta}^{\gamma} \bar{\bar{\omega}}_{\gamma} \\ \bar{W}_3^{\alpha} b_{\alpha}^{\beta} \bar{\bar{I}}_{\beta}^{\gamma} \bar{\bar{\omega}}_{\gamma} \end{pmatrix} \tag{4.71}$$

where $\bar{\boldsymbol{r}}_i$ defines the rotating vector to the center of pressure point from the rotation axis of the propeller in body axis coordinates. The second term in the summation is the pitching moment of the propeller blade itself, as shown in Fig. 4.5. The last terms of Eq. (4.71) are the vector components of the body axis torques from Euler's rotational equation given by Eq. (3.77). These torques account for the body axis torques due to the angular momentum of the rapidly rotating engine and propeller.

The instantaneous required torque to drive the propeller shaft is

$$\bar{N}_{\mathrm{REQ}} = \sum_{i=1}^{N} (-\bar{C}_{L_i} \cos\alpha_{\mathrm{sec}} - \bar{C}_{D_i} \sin\alpha_{\mathrm{sec}}) \bar{q}_p \bar{S} \bar{r} = \sum_{i=1}^{N} \bar{f}_{a_i} \bar{r} \tag{4.72}$$

where $\bar{r}$ is the radius to where the simulated propeller lift and drag forces are acting. Notice that from Eqs. (4.68), (4.71), and (4.72), although the airplane flies at subsonic speeds, the propeller blade Mach numbers can be high and can approach sonic speeds. For example, for a 72-in. diam propeller operating at 2500 rpm with the airplane flight speed of 100 nmph, the propeller tip Mach number is 0.72. Hence, estimations for $\bar{C}_{L_i}$, $\bar{C}_{D_i}$, and $\bar{C}_{m_i}$ in Eqs. (4.68), (4.71), and (4.72) must include the effects due to compressibility (Ref. 56, Sec. 11.4), i.e., Mach number effects.

The torque to drive the propeller is supplied by the power ***available*** from the airplane engine that can provide the following power (in ft-lb/s) to the propeller shaft:

$$P_{AVAL} = 550\eta(THROT)(HP_{\mathrm{MAX}}) \tag{4.73}$$

where η is the mechanical efficiency of the engine and generally has a value of about 0.80. *THROT* is the throttle position, which will be between 0.0 and 1.0 and is controlled by the pilot or a model of the pilot's actions during the simulation. The throttle position in turn controls the air intake *manifold pressure* to the engine. The power being developed by the reciprocating engine is a function of the pressure in the engine intake manifold. Estimates for the torque available from the engine to drive the propeller shaft is simply $\bar{N}_{AVAL} = P_{AVAL}/\bar{\bar{\omega}}_r$, where again $\bar{\bar{\omega}}_r$ is the engine and propeller rotational speed in rad/s. The differential equation of motion for the engine and propeller rpm is approximated by

$$\bar{N}_{AVAL} - \bar{N}_{\text{REQ}} = \bar{\bar{I}}_1^1 \dot{\bar{\omega}}_r \tag{4.74}$$

where $\bar{\bar{I}}_1^1$ is the combined moments of inertia of the rotating engine parts and the propeller. Numerical solutions for Eq. (4.74) may require cycle times between 100 and 200 Hz to simulate the rotational speed of the propeller and approximations for the thrust. If the simulated propeller is a *constant speed propeller*, the propeller angle of incidence (the blade angle), $i_{ba} = i_{\text{sec}} + i_{\text{prop}}$, is modulated (constantly being changed by i_{prop}) by the propeller governor to control the engine rpm, or $\bar{\omega}_r$ in rad/s. With a constant speed propeller, the pilot can control the engine rpm by positioning the propeller control lever in the cockpit. Hence, Eq. (4.74) is constantly forced to zero by changing the propeller angle of incidence to achieve the desired engine rpm, thus eliminating rpm as a variable in the analysis. Thrust computations are a part of the complete airplane motion simulation, for example, as described in Sec. 2.2.3.

The specific fuel consumption f_s of the reciprocating piston engine is generally given in lb/h/hp. The weight of the airframe as a function of time in hours can be estimated by

$$w(t) = w(t_o) - \int_{t_o}^{t} (THROT)(HP_{\text{max}}) f_s \, \mathrm{d}t \tag{4.75}$$

where *THROT* is again the instantaneous throttle setting, and HP_{max} is the maximum rated horsepower of the engine.

4.2.6 Simulating Flight Vehicle Engine Thrust: The Airbreathing Jet Engine

This is a general discussion of the simulation equations used for turbojet engine modeling. These equations can include the momentum changes due to the geometric effects of nonparallel inlet and exhaust directions as well as the fluid pressure forces acting over the inlet and exhaust areas.

The thrust vector created by the turbojet engine (Ref. 34, Secs. 9.4, 9.5) is given by

$$\boldsymbol{T} = (\dot{w}_{\text{air}} + \dot{w}_{\text{fuel}})\boldsymbol{V}_e - \dot{w}_{\text{air}}\boldsymbol{V} + (p_e - p_o)\boldsymbol{A}_e - (p_i - p_o)\boldsymbol{A}_i \tag{4.76}$$

The $\dot{w}_{\text{air}}$ and $\dot{w}_{\text{fuel}}$ are the weight flow rates in slugs/s, of air and fuel passing through the engine, respectively, p_o is the ambient air pressure, and p_i and p_e are the inlet and exhaust pressures, respectively. The area vectors are formed from the outward pointing normals to the inlet and exhaust areas, i.e., $\boldsymbol{A}_i$ and $\boldsymbol{A}_e$, respectively.

An imaginary turbojet engine configuration is visualized in Fig. 4.8. The inlet static air pressure p_i, is usually very close to atmospheric pressure. For maximum thrust, the exhaust nozzle can be designed to allow the exhaust gases to be further

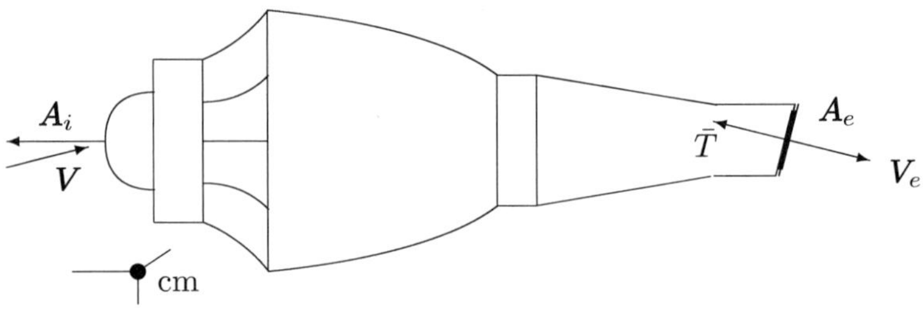

Fig. 4.8 Thrust forces acting on the turbo-jet engine.

accelerated by expanding the gas in the tailpipe so that at the nozzle exit, we have $p_e = p_o$. Ideally, all pressure energy produced in the engine will be converted to the kinetic energy of the exhaust gases, whereby producing a maximum exhaust velocity, V_e and hence, maximizing the engine thrust. If the jet exhaust and inlet velocity directions are nearly parallel, then the thrust magnitude from Eq. (4.76) can be simplified to

$$\bar{T}_1 = \dot{w}_{\text{air}}(V_e - V) + (p_e - p_o)A_e \tag{4.77}$$

since the ratio $\dot{w}_{\text{fuel}}/\dot{w}_{\text{air}}$ is generally less than 0.05 and is sometimes neglected. The effects of inlet pressure forces may be included in the estimates for the aerodynamic derivatives by adding small deltas to the force and moment coefficients, which would be functions of engine power settings. However, the thrust from the exhaust gas of the jet engine is large compared to the weight of the airplane, and its body axis dynamic effects can be estimated. For example, when the engine exhaust direction is vectored and like Eq. (4.69) can be defined by two rotation angles, θ_1 and θ_2, relative to the airplane body axis, the thrust vector in airplane body axis coordinates is again given by

$$\bar{\boldsymbol{T}} = \bar{T}_1 \begin{pmatrix} \cos\theta_1 \cos\theta_2 \\ \sin\theta_1 \cos\theta_2 \\ -\sin\theta_2 \end{pmatrix} \tag{4.78}$$

The airplane body axis moments due to jet engine thrust can be approximated with

$$\begin{pmatrix} \mathcal{L} \\ \mathcal{M} \\ \mathcal{N} \end{pmatrix} = \boldsymbol{d} \times \bar{\boldsymbol{T}} - \begin{pmatrix} \bar{W}_1^{\alpha} b_{\alpha}^{\beta} \bar{\bar{I}}_{\beta}^{\gamma} \bar{\bar{\omega}}_{\gamma} \\ \bar{W}_2^{\alpha} b_{\alpha}^{\beta} \bar{\bar{I}}_{\beta}^{\gamma} \bar{\bar{\omega}}_{\gamma} \\ \bar{W}_3^{\alpha} b_{\alpha}^{\beta} \bar{\bar{I}}_{\beta}^{\gamma} \bar{\bar{\omega}}_{\gamma} \end{pmatrix} \tag{4.79}$$

where the vector $\boldsymbol{d}$ is from the airplane's center of mass to the center of the exhaust gas tailpipe outlet. The last terms in Eq. (4.79) are the body axis moments due to the angular momentum of the rapidly rotating jet engine parts, as given in Eq. (3.77). For steady-state flight attitudes, i.e., when the airplane's body axis rotation rates are zero, these torques on the airframe will also be zero. However, these gyroscopic torques can be high, especially whenever the aircraft is rapidly rotated. For example, these torques are high when the aircraft is rotated to the flight attitude during the takeoff roll, i.e., when there is a sudden increase in the airplane pitch angle by increasing q, the pitch rate. The body axis moment coefficient's contributions from the jet engine thrust, as shown in Eq. (4.44), become

$$\begin{pmatrix} C_l \\ C_m \\ C_n \end{pmatrix} = \left(\frac{1}{\bar{q}S}\right) \begin{pmatrix} \mathcal{L}/b \\ \mathcal{M}/\bar{c} \\ \mathcal{N}/b \end{pmatrix} \tag{4.80}$$

The power output of the jet engine is simply controlled by the amount of fuel being burned by the engine. The pilot's throttle setting, *THROT*, controls fuel to the engine and thus controls the engine power within the operating range of the engine. In a similar manner to Eq. (4.75), the airframe weight as a function of time can be approximated by

$$w(t) = w(t_o) - \int_{t_o}^{t} (THROT) T_{\max} f_s \, \mathrm{d}t \tag{4.81}$$

where $T_{\max}$ is the maximum engine thrust available, and f_s is the specific fuel consumption in lb/h per pound of thrust. The fuel flow rate, f_s, is a sensitive function of the atmospheric parameters, such as temperature, pressure, and altitude (density).

4.2.7 Simulating Flight Vehicle Engine Thrust: The Rocket Engine

The following discussion is limited to rocket engines using the heat energy from the combustion of chemical propellants. Other forms of energy generation, such as nuclear and ion-type engines, are beyond the scope of this book. However, these other forms of energy can very well indeed be the forms of propulsion for the future that can take mankind to much greater distances into space and within human time limitations.

The thrust force magnitude for each chemical propellant rocket engine mounted on the space vehicle is given by

$$\bar{\bar{T}}_1 = \dot{w} I_{\mathrm{sp}} + (p_e - p_o) A_e \tag{4.82}$$

where the force direction is opposite to the exhaust gases being ejected from the engine or along the nozzle's longitudal axis. The $\dot{w}$ is the weight flow rate in lb/s, and the I_{sp} is the specific impulse performance of the ideal rocket engine and its propellant combination. If the exhaust gas pressure is expanded through the nozzle to exit with $p_e = p_o$, we have the ideal rocket equation as presented in Sec. 2.1.4. A typical rocket engine configuration, with the defining parameters, can be described using Fig. 4.9. The exhaust velocity magnitude is given by $V_e = g_o I_{\mathrm{sp}}$. When the rocket engines are large, they may be mounted on movable gimbals so that the angles θ_1 and θ_2 are controlled to provide the necessary torques about the vehicle's axes for attitude control and to maintain the optimum thrust direction. This control action is much the same as balancing a broom vertically on the end

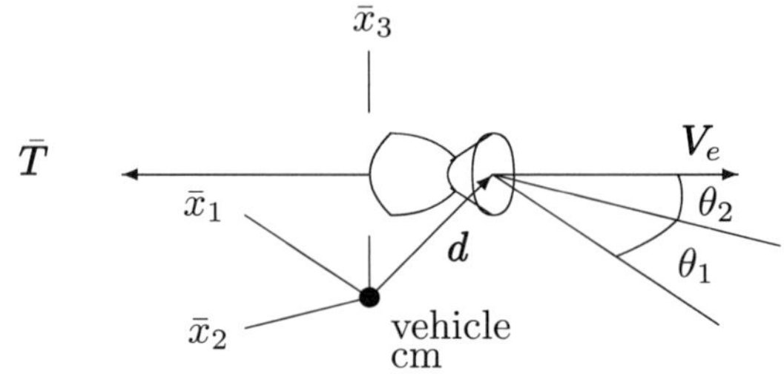

Fig. 4.9 Thrust forces acting on the rocket engine.

of your finger. On the other hand, when the rocket engines are small and are used primarily for vehicle attitude control, the angles θ_1 and θ_2 are fixed, thus providing known amounts of body axis torques each time an engine is fired. Based on the angles similar to those in Eq. (4.69), each rocket engine thrust vector in the body axis coordinate frame becomes

$$\bar{\boldsymbol{T}} = \bar{\bar{T}}_1 \begin{pmatrix} \cos(\pi - \theta_1)\cos\theta_2 \\ \sin(\pi - \theta_1)\cos\theta_2 \\ -\sin\theta_2 \end{pmatrix} \tag{4.83}$$

Most rocket engines are not throttlable and are designed to operate at maximum thrust, so that the propellant weight flow rate is given by

$$\dot{w} = \frac{\bar{T}_{\text{ideal}}}{I_{\text{sp}}} \tag{4.84}$$

where $\bar{T}_{\text{ideal}}$ is the ideal rocket engine thrust of the engine in a vacuum, i.e., there are no pressure losses $(p_e - p_o) = 0$, from Eq. (4.82). The weight of the space vehicle as a function of time in seconds simply becomes

$$w(t) = w(t_o) - \dot{w}(t - t_o) \tag{4.85}$$

The rocket engine thrust force becomes part of the total contact force acting on the center of mass of the space vehicle, causing the translational accelerations.

The instantaneous moment for each rocket engine about the flight vehicle's center of mass becomes

$$\boldsymbol{N} = \boldsymbol{d} \times \bar{\boldsymbol{T}} \tag{4.86}$$

These moments are used to steer the space vehicle's attitude in inertial space and are the controlling torques used in the solution of Euler's equation for the rotational dynamics of the space vehicle, as presented in Sec. 3.2.5 and given in Eq. (3.82).

The chemical performance of the rocket engine is given by the specific impulse I_{sp}. During translational maneuvers, the actual rocket vehicle propulsion system performance is measured by the amount of energy change in the space vehicle's trajectory in inertial space. The energy may be added to or subtracted from the trajectory. For instance, energy is added to the vehicle during ascent to orbit and subtracted from the orbit during a de-orbit burn. The integrals of the inertial accelerations from engine thrust and other contact forces form the total delta-V (delta velocity) gained by the space vehicle. For example, the total delta-V from a particular rocket engine burn is

$$\Delta V_T = \int_{t_{ig}}^{t_{bo}} \left| (g\boldsymbol{T}/w) - (C_D\bar{q}S/w)\left(\frac{\boldsymbol{V}}{|\boldsymbol{V}|}\right) \right| \mathrm{d}t \tag{4.87}$$

where T, w, and $\bar{q}$ are functions of time; t_{ig} and t_{bo} are the rocket engine ignition time and the burnout time, respectively. The first term in the integral is the ΔV from the rocket engine thrust forces, and the second term is the ΔV losses from aerodynamic drag forces, which may approach zero for on-orbit operations.

The integrands in Eq. (4.87) can be accurately estimated from the vehicle's onboard accelerometers, which are mounted on the IMU stable member platform, as described in Sec. 1.3.6. These accelerometers will detect the accelerations due to all of the contact forces acting on the space vehicle. The inertial acceleration vector can be determined from the *inertial-to-stable-member* transformation given by Eq. (1.97). Estimates for the components of the inertial contact acceleration vector, calling it $\dot{\boldsymbol{v}}_c$, from the IMU measurements are simply

$$\hat{\ddot{x}}^i = R^i_\alpha \ddot{y}^\alpha \tag{4.88}$$

The estimate for the total acceleration magnitude becomes

$$|\dot{\boldsymbol{v}}_c| = \hat{a}_\tau = \sqrt{\hat{\ddot{x}}_1^2 + \hat{\ddot{x}}_2^2 + \hat{\ddot{x}}_3^2} \tag{4.89}$$

and Eq. (4.87) is estimated by

$$\Delta \hat{V}_T = \int_{t_{tig}}^{t_{bo}} \hat{a}_\tau \, \mathrm{d}t \tag{4.90}$$

Notice that the total ΔV given in Eq. (4.87) is independent of the direction of the applied contact thrust force. However, to make efficient use of the transfer of energy to the flight vehicle's translational and/or rotational motion, control of the rocket thrust direction in inertial space is necessary and very important to fly the vehicle along the desired trajectory. For example, it can be shown that the maximum energy transfer to the flight vehicle's orbit can be achieved by controlling the *thrust direction to always be aligned with the instantaneous inertial velocity vector*. Here, the term *gravity turn* is used to describe a burn maneuver that is a segment of a powered flight trajectory. The term *gravity turn maneuver* is also used to describe the changing or turning of a space vehicle's planetary trajectory by passing close to another planet or satellite. In this context, no rocket thrust is expended, and the gravitational acceleration of the planet or satellite is used to target the space vehicle along the desired flight path.

The ΔV in the direction of the inertial velocity direction, $\Delta \hat{V}_V$, is one of the *burn performance parameters* used to control the rocket engine thrust direction to

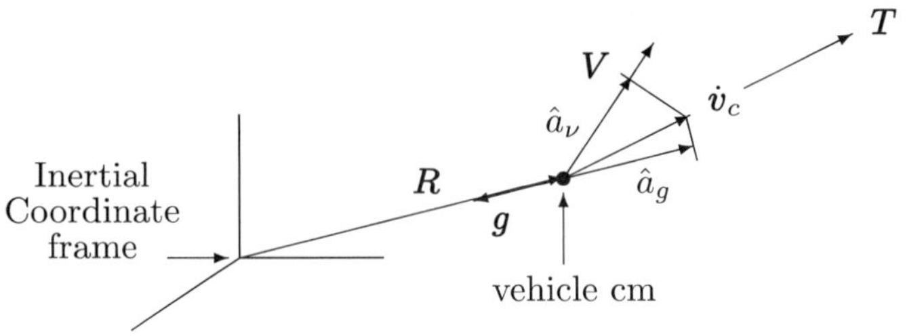

Fig. 4.10 Contact accelerations parallel and perpendicular to the local inertial velocity and gravity directions.

optimize the transfer of energy to the space vehicle's orbit. The differential burn performance parameters can be visualized in Fig. 4.10.

In the GNC portion of the flight software as discussed in Sec. 1.4.8, the instantaneous magnitude of the contact accelerations parallel to the $\boldsymbol{V}$ direction, as shown in Fig. 4.10, is estimated by

$$\hat{a}_\nu = \frac{(\boldsymbol{V} \cdot \dot{\boldsymbol{v}}_c)}{|\boldsymbol{V}|} \tag{4.91}$$

The accumulated velocity in the $\boldsymbol{V}$ direction becomes

$$\Delta \hat{V}_V = \int_{t_{\text{tig}}}^{t_{\text{bo}}} \hat{a}_\nu \, \mathrm{d}t \tag{4.92}$$

In the same manner, the accumulated velocity due to contact accelerations parallel to the $\boldsymbol{g}$ direction, also shown in Fig. 4.10, becomes

$$\Delta \hat{V}_g = \int_{t_{\text{tig}}}^{t_{\text{bo}}} \hat{a}_g \, \mathrm{d}t \tag{4.93}$$

This burn performance parameter is a measure of the energy lost from accelerating the mass of the space vehicle against the acceleration due to gravity. This is called the *gravity loss term* or simply *gravitation losses* and is the burn performance parameter to be minimized by the GNC functions, especially during the powered ascent to orbit. The burn performance parameters can become part of the integral list in the onboard navigation software as shown in Appendix C.3.

Launching from the Earth's surface, the space vehicle cannot initially be accelerated horizontally, but must rise vertically to avoid the buildup of speed in the denser parts of the atmosphere. Hence, gravitational losses are unavoidable. As the altitude increases, the space vehicle is pitched over (thrusting more horizontally), and more $\Delta \hat{V}_V$ can be added to the orbit.

When the dynamic pressure $\bar{q}$ becomes low enough, say above 50 n miles altitude, the space vehicle is pitched nearly horizontal, and a gravity turn maneuver can be performed to complete the acceleration of the vehicle to the desired perigee velocity for orbital injection. Some gravity losses can be recovered by *lofting* the trajectory to a slightly higher altitude and then allowing the altitude to decrease as the vehicle approaches the rocket engine cutoff time. The optimized trajectory for ascent from the surface of Earth to orbit was first done by Hermann Oberth in 1925 (Ref. 58, p. 300) and was called the *Oberth synergy curve.*

Launching from the surface of an airless body, such as the lunar surface, gravitational losses can be minimized by orbital injections at very low altitudes. The desired ascent trajectory can be summarized as follows: The space vehicle lifts off vertically and initially accumulates gravity losses. The altitude rise (again lofting the trajectory) necessary is *that height required such that when a gravity turn maneuver is performed, the vehicle will reach orbit velocity at the predetermined lower altitude.* Hence, nearly all of the initial gravity losses are recovered during the fall to the lower altitude (higher gravitational potential) just prior to reaching the desired orbital injection velocity. For an excellent and detailed discussion of the preceding powered flight trajectories, see Ref. 22 (Chap. 6).

4.2.8 Solar Radiation Pressure Forces

A force due to solar radiation is produced when flight vehicle surface areas are exposed to light from the sun. This force comes from the exchange of momentum (Ref. 8, Sec. 8.6.4) as the light particles, *photons*, strike and are reflected off the vehicle surface areas. Since photons travel at the speed of light, the total energy of these particles is given by Einstein's mass-energy relationship (Ref. 1, p. 103), $E_s = mc^2$, and the total momentum of the particles simply becomes

$$mc = \frac{E_s}{c} \tag{4.94}$$

where c is the speed of light. These forces are generally quite small and are dwarfed at lower altitudes by atmospheric drag forces and the gravitational gradient torques. However, at higher altitudes and for space vehicles with large area-to-mass ratios, the total force becomes a significant perturbation to the orbital motion and can also affect rotational motion about the center of mass. These effects on the orbital motion accumulate during long periods of flight.

The solar energy constant (the energy of the stream of photons coming from the sun) used for engineering applications is about $E_s = 1353\,\mathrm{W/m^2}$ (Ref. 31, p. 225) at a distance of one *astronomical unit*. The astronomical unit (AU) (Ref. 5, p. M2) is the unit of measure slightly less than the semimajor axis of the Earth's orbit around the sun, 1 AU is $\mathrm{A} = 1.49597870691 \times 10^{11}$ m (Ref. 5, p. K6) or 92,955,807.21 statute miles, the more familiar number of about 93 million miles.

Continuing with Eq. (4.94), we have the solar radiation pressure constant, calling it $\bar{q}_s$, $4.510 \times 10^{-6}\,\mathrm{N/m^2}$, and in English units $9.419 \times 10^{-8}\,\mathrm{lb/ft^2}$. Analogous to the fluid dynamic pressure $\bar{q}$ used in the computation of aerodynamic forces as described in Sec. 2.3.3 and again in Sec. 4.2.2, we can estimate the solar radiation force on the ϱth finite area of the space vehicle with

$$\bar{\boldsymbol{f}}_\varrho = -C_\varrho \bar{q}_s S_\varrho (\bar{\boldsymbol{u}}_s \cdot \bar{\boldsymbol{n}}_\varrho)\bar{\boldsymbol{n}}_\varrho \tag{4.95}$$

where $\bar{\boldsymbol{f}}_\varrho$ is the body axis force on the ϱth finite area S_ϱ, which is the exposed surface area with $\bar{\boldsymbol{n}}_\varrho$ as its outward pointing unit normal. The dot product term is the cosine of the angle between the stream of photons and the exposed panel area. This adjusts the total area exposed to the stream of photons based on the panel orientation. C_ϱ is the force coefficient for that panel area, which will be discussed in some detail in the next paragraphs.

The $\bar{\boldsymbol{u}}_s$ is the body axis unit pointing vector to the sun. This vector can be derived from the sun pointing inertial vector as shown and defined in Fig. 3.2:

$$\boldsymbol{R}_{VS} = \boldsymbol{R}_{ES} - \boldsymbol{R}_{EV} \tag{4.96}$$

The distance from the vehicle to the sun becomes $R_{VS} = |\boldsymbol{R}_{VS}|$, and in AU, $R_{\text{AU}} = R_{VS}/\text{A}$, whereas A is the astronomical unit as defined above. The unit vector to the sun in the inertial frame is $\boldsymbol{u}_{VS} = \boldsymbol{R}_{VS}/R_{VS}$. From the navigation software or the onboard GNC software, we will have the body-axis-to-inertial transformation matrix a, and we have

$$\bar{\boldsymbol{u}}_s = (a)^T \boldsymbol{u}_{VS} \tag{4.97}$$

which is the instantaneous sun pointing vector in the body axis coordinate frame. Equation (4.95) is the solar radiation force on a finite area panel of a space vehicle, which is *located at 1 AU from the sun*. For the space vehicle located at R_{AU} distance from the sun, Eq. (4.95) becomes

$$\bar{\boldsymbol{f}}_\varrho = -C_\varrho S_\varrho (\bar{\boldsymbol{u}}_s \cdot \bar{\boldsymbol{n}}_\varrho) \left(\frac{\bar{q}_s}{R_{\text{AU}}^2} \right) \bar{\boldsymbol{n}}_\varrho \tag{4.98}$$

This solar radiation pressure force perpendicular to the panel area can be visualized using Fig. 4.11. The momentum actually transferred to the finite area S_ϱ is a function of its surface *reflectivity*. The force coefficient C_ϱ becomes the coefficient of reflectivity C_{r_ϱ}, which is determined by the surface material of each panel area. Panel color, transparency, temperature, and even its texture determine the surface reflectivity. Examples of surface reflectivity values can be summarized as follows:

1) $C_{r_\varrho} = 0.0$: like clear glass, where the surface is transparent; no momentum is transferred.

2) $C_{r_\varrho} = 1.0$: a black body, where all of the radiation is absorbed on initial surface contact, i.e., in Fig. 4.11, the reflected photon stream $\bar{q}_r = 0.0$.

3) $C_{r_\varrho} = 2.0$: like a mirror, where the momentum of the incoming radiation is first transferred and then reflected back into space so that $\bar{q}_r = (\bar{q}_s/R_{\text{AU}}^2)$.

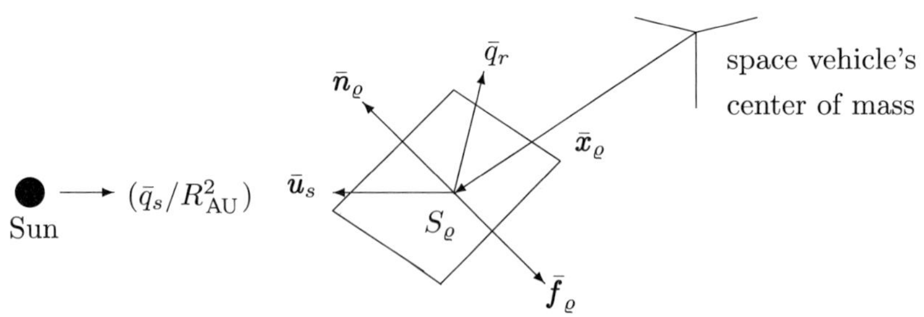

Fig. 4.11 Solar radiation pressure acting on a finite element area panel of the space vehicle.

The determination of the panel surface area coefficients of reflectivity is complex, and one can see that with C_{r_ϱ} values around one, that small sheer forces (forces parallel to the panel surface areas analogous to viscous forces in aerodynamic fluid flow) can be transferred to the panel area. These sheer forces are suggested in Ref. 8, (p. 545) and in the literature, but a discussion here is beyond the scope of this text. Nevertheless, Eq. (4.98), with $C_\varrho = C_{r_\varrho}$, can be used as a first approximation for the solar radiation pressure force on each finite panel area.

When estimating the solar radiation pressure perturbations to the motion of the space vehicle, two additional factors must be included in the computational logic: 1) the entire space vehicle may be shadowed by an astronomical body, such as the Earth, in which case $\bar{q}_s$ will approach zero, and 2) some finite panel areas of the space vehicle may be shadowed by other panel areas, i.e., by its own antennas or other structural members. Hence, for each of the ϱth panel area force estimates, from Eq. (4.98), all other panel locations must be examined for blocking the sunlight onto the ith panel area. When the sunlight on the ith panel is blocked, then $\bar{\boldsymbol{f}}_\varrho = 0.0$.

When the forces on all panel areas, again using n_ϱ, for the number of panels representing the space vehicle, we can compute the total force in body axis coordinates due to solar radiation pressure:

$$\bar{\boldsymbol{F}} = \sum_{\varrho=1}^{n_\varrho} \bar{\boldsymbol{f}}_\varrho \tag{4.99}$$

The perturbing *inertial acceleration* due to solar radiation pressure $\dot{\boldsymbol{v}}_{sp}$, which affects the orbital motion of the space vehicle, can also be expressed as follows:

$$\dot{\boldsymbol{v}}_{sp} \approx -\left(\frac{\tilde{C}S_{\text{ref}}}{m}\right)\left(\frac{\bar{q}_s}{R_{\text{AU}}^2}\right)\boldsymbol{u}_{VS} \tag{4.100}$$

where m is the mass of the space vehicle, and S_{ref} is an effective reference area for the space vehicle. Notice that the perturbing acceleration is assumed to be directed away from the sun pointing vector. The term $(\tilde{C}S_{\text{ref}})/m$ can become a solved-for parameter in the orbit determination processors as described in Sec. 4.1.3. Initial values for $\tilde{C}$ can be estimated from the finite element area computational results of Eq. (4.100) as follows: If the space vehicle is stabilized to maintain a constant attitude relative to the sun pointing vector, then the total body solar radiation pressure coefficient can be estimated with

$$\tilde{C} \approx \left(\frac{R_{\text{AU}}^2}{\bar{q}_s S_{\text{ref}}}\right)|\bar{\boldsymbol{F}}| \tag{4.101}$$

The space vehicle may be designed to use the body axis torques from the solar radiation pressure forces as a weak control system to hold a sun relative stabilized attitude. This can be done by positioning the space vehicle's solar panel structures on relatively long moment arms from the vehicle center of mass, thereby increasing the torques create by the solar radiation forces.

Control panel areas on the extended arms can be designed to rotate and control the angle between the sun and the unit normal to the panel ($\bar{\boldsymbol{u}}_s \cdot \bar{\boldsymbol{n}}_\varrho$), from Eq. (4.98). In this manner, control torques can be used to maintain the attitude of the space vehicle relative to the sun pointing vector. These body axis torques, which are used in the solution of Euler's rotational equation of motion given in Eq. (3.82), can control the rotational dynamics of the space vehicle. From the finite element area analysis, the body axis torques due to solar radiation pressure become

$$\bar{\boldsymbol{N}}_s = \sum_{\varrho=1}^{n_\varrho} (\bar{\boldsymbol{x}}_\varrho \times \bar{\boldsymbol{f}}_\varrho) \tag{4.102}$$

4.2.9 Estimating On-orbit Aerodynamic Drag Effects

The on-orbit drag force acting on a space vehicle due to the upper levels of the atmosphere cause a loss of the orbit energy and a decay of the orbit. The drag force is the result of air particles impinging on the space vehicle's surface areas, which are exposed to a slipstream with Mach numbers in excess of 20. At Earth orbital speeds and altitudes, the atmospheric particles simply bounce off the space vehicle's surfaces and rebound absorbing about the same amount of energy as they had as they hit the surface. This flow region of aerodynamics is called *molecular flow*, as outlined in Sec. 4.2.3. In this region, the aerodynamics become somewhat simpler, and the drag coefficients approach a value of 2. Figure 4.12 depicts the methods used to estimate the pressure force on a finite element area of the space vehicle's exposed surface area.

The dynamic pressure at orbiting altitudes is dependent on the estimates for the density ρ of the oncoming atmospheric particles. The density at orbiting altitudes is very sensitive to solar activity, as given by the value of the solar flux and the value of the geomagnetic index, as described in Sec. 4.2.1. For example, during periods of high solar activity, the density can double in a short period of time as more protons, neutrons, or electrons enter the Earth's upper atmosphere.

Again using finite area analysis as in Secs. 4.2.3 and 4.2.8, the instantaneous drag force on the ϱth panel, which is similar to Eq. (4.48) but is a function of the

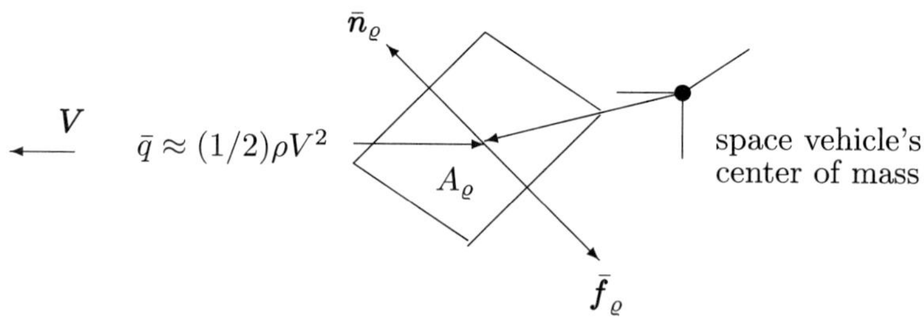

Fig. 4.12 Fluid dynamic pressure acting on a finite element area panel of the space vehicle.

angle that the slip stream impinges on the finite panel area, can be approximated as follows:

$$\bar{f}_{\varrho} \approx -C_{D_{\varrho}}\bar{q}A_{\varrho}\left[(\bar{n}_{\varrho})\cdot\left(\frac{V}{|V|}\right)\right]\bar{n}_{\varrho} \tag{4.103}$$

where the force from the aerodynamic pressure is opposite the direction of the outward pointing unit normal to the panel area. The total body axis force vector due to the atmospheric density becomes

$$\bar{F} = \sum_{\varrho=1}^{n_{\varrho}} \bar{f}_{\varrho} \tag{4.104}$$

The body axis moments from aerodynamic forces become

$$\bar{N}_a = \sum_{\varrho=1}^{n_{\varrho}} (\bar{x}_{\varrho} \times \bar{f}_{\varrho}) \tag{4.105}$$

We can also approximate the total drag force acting on the space vehicle's center of mass, as shown with Eq. (2.117), as follows:

$$D \approx -C_D\bar{q}S\left(\frac{V}{|V|}\right) \tag{4.106}$$

Using the standard definition for a drag force, the total drag force is in the opposite direction of the relative *wind vector*. The drag coefficient is dependent on the space vehicle's wind relative attitude, i.e., $C_D = C_D(\alpha, \beta)$, where α is the angle of attack and β is the sideslip angle. If the attitude of the space vehicle relative to the inertial velocity vector is held constant, i.e., α and β are constant, then the total force magnitude in Eq. (4.103) can be used to estimate the total drag force in Eq. (4.104). Using a reference frontal area S for the space vehicle, we can write

$$|\bar{F}| \approx C_D\bar{q}S \tag{4.107}$$

This constant space vehicle attitude is called an *LVLH hold attitude*. In this attitude control mode the space vehicle should be designed to be near a *torque equilibrium attitude*. This attitude is simply where the sum of all body axis torques are near zero, and the space vehicle's attitude would stabilize and remain in this attitude equilibrium condition. Using Euler's rotational equation of motion given by Eq. (3.82), we can write the LVLH relative function to be minimized as

$$f(\psi,\theta,\phi) = \bar{N}_{g\alpha} + \bar{N}_{a_{\alpha}} + \bar{N}_{s_{\alpha}} - \bar{W}_{\alpha}^{\beta}\bar{I}_{\beta}^{\gamma}\bar{\omega}_{\gamma} \tag{4.108}$$

where ψ, θ, and ϕ are the yaw-pitch-roll Euler angles orienting the space vehicle body axis to the LVLH coordinated frame. For an LVLH hold attitude, the body axis rotation rate vector $\bar{\omega}$ is $(0, -\omega_o, 0)$, where ω_o is the LVLH coordinate axis rotation rate relative to the inertial frame given by Eq. (2.141) and is called the *orbital rate*. The $\bar{N}_{g_\alpha}$ are vector components of the gravity gradient torques given by Eq. (3.81), the $\bar{N}_{a_\alpha}$ are the aerodynmaic torques from Eq. (4.105), and the $\bar{N}_{s_\alpha}$ are the solar radiation torques from Eq. (4.102), which are generally quite small and neglected for low Earth orbiting vehicles.

The space vehicle design configuration is very important so that 1) the space vehicle will maintain this attitude with a minimum expenditure of control torques and 2) if no control torques are available, i.e., if power is off on the space vehicle, the vehicle will maintain a fixed LVLH attitude and not tumble out of control.

The drag contact force causes a drag acceleration that perturbs the space vehicle's orbit. For a space vehicle of mass m, the inertial acceleration becomes

$$\boldsymbol{a}_D = -\left(\frac{C_D S}{W}\right) g_o \bar{q} \left(\frac{\boldsymbol{V}}{|\boldsymbol{V}|}\right) \tag{4.109}$$

where the weight of the vehicle is $W = mg_o$, and g_o is the gravity constant, $32.174048556\,\mathrm{ft/s^2}$. The term $B = (C_D S/W)$ is sometimes called the vehicle's ballistic coefficient. Note that different authors may define the ballistic coefficient with different groupings of the space vehicle's body-dependent parameters, C_D, S, and m.

Using the differential energy theorem given by Eq. (2.14), the rate of change of the orbital energy due to the perturbing aerodynamic drag force is

$$\frac{\mathrm{d}E}{\mathrm{d}t} = -D|\boldsymbol{V}| \tag{4.110}$$

The Keplerian total energy of the space vehicle in orbit is given by Eq. (2.58). Differentiating this equation with respect to time, we have

$$\frac{\mathrm{d}E}{\mathrm{d}t} = \frac{\mu}{(2a^2)} m \frac{\mathrm{d}a}{\mathrm{d}t} \tag{4.111}$$

Most orbital operations are designed to fly in near-circular orbits for the simplicity of many GNC functions and to accomplish the space vehicle's mission requirements. Using Eq. (2.60) for the circular orbit speed and equating Eqs. (4.110) and (4.111), we have the first approximation for drag decay of the orbit's semimajor axis:

$$\frac{\mathrm{d}a}{\mathrm{d}t} \approx -2B(g_o\bar{q})\left(\frac{a}{V}\right) \tag{4.112}$$

This relationship is useful because it estimates the orbit decay rate using its three distinct functional dependents. 1) The space vehicle structural configuration defined by B, the ballistic coefficient; 2) the atmospheric parameter $\bar{q}$, the dynamic pressure; and 3) the orbit-specific parameters, the orbit semimajor axis a and the orbital speed V.

4.2.10 Aerospace Applications: Six-Degrees-of-Freedom Airframe Motion Simulation

This section presents a brief outline for computing the aerodynamic forces and moments that can be used to describe the six-DOF motion of a conventional wing-tail airplane in subsonic flight. The analysis is intended to model methods that could be used in high-speed computer systems to drive aircraft flight simulators or to study aircraft flight and stability characteristics. There are many very good basic texts, such as Ref. 35, (Chap. 5) and Ref. 34 (Chap. 7), which present this analysis in detail and are the basis for this computational method. This method adds to the traditional computational methods by using coordinate transformations to estimate the relative geometric air flow angles to the lifting surfaces.

Using the definitions for the aerodynamic lifting surfaces as presented in Sec. 4.2.4, Fig. 4.13 depicts a simplified airplane configuration showing the forces on the airframe and basic design parameters in the $\bar{x}_1$-$\bar{x}_3$ plane, i.e., a view from the left side of the airplane. The thrust vector is skewed somewhat, as suggested in Sec. 4.2.5, here to emphasize that the propellor thrust may not be parallel to the airframe body axes. Notice that for this application we start with the unbarred frame representing the Earth-fixed station coordinate frame and the barred frame is selected to represent the body axis coordinate frame. Also note that the airflow angles are shown in body axis coordinates for demonstration purposes.

Using Fig. 4.13, we have the following definitions:

1) α, the airplane angle of attack,

2) $\bar{q}_w$, the wing lifting surface dynamic pressure,

3) α_w, the wing lifting surface angle of attack, and

4) $\Delta\alpha$, the downwash flow angle due to the airplane wing lifting surfaces Ref. 35, p. 222). C_L is the *airplane wing lift coefficient*, and AR is the airplane wing aspect ratio.

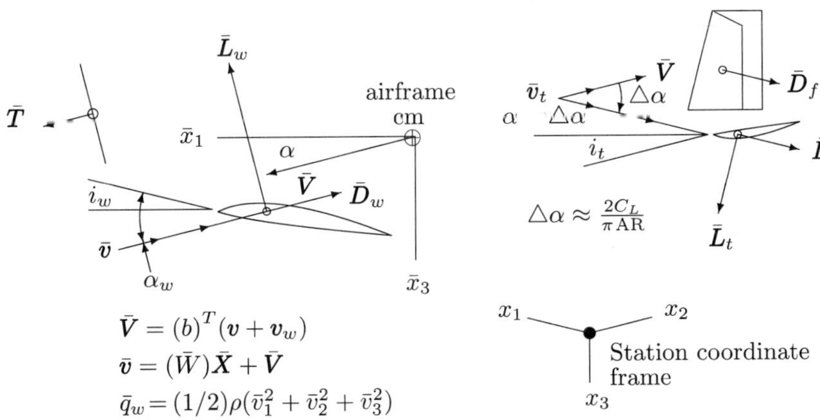

Fig. 4.13 Parameters defining the geometry of the airplane wing and tail configuration.

Part 1: Computations for the dynamic parameters from the aircraft state vector and atmospheric winds. This part is similar to Part 1 as described in Sec. 2.3.3; however, here we will require much more detailed analysis. The aircraft state vector for the six-DOF simulations becomes $(t, \boldsymbol{x}, \boldsymbol{v}, \psi, \theta, \phi, p, q, r)$, where $\boldsymbol{x}$ and $\boldsymbol{v}$, for this analysis, are the position and velocity vectors of the aircraft's center of mass in *station coordinates*, respectively. The airplane yaw ψ, pitch θ, and roll ϕ angles define the body-axis-to-station-coordinate transformation matrix, $(b) = Z(\psi)Y(\theta)X(\phi)$. The body axis rotation rate components are p, q, and r. The control parameters for this simplified example might be δ_e, the elevator control effector deflection angle; δ_r, the rudder control effector deflection angle; δ_a, the aileron control effector deflection angle; and δ_t, the thrust control effector.

The geometry of the aircraft configuration can be represented in a database by defining the ϱth lifting surface as follows:

1) $\bar{X}_{\varrho 1}, \bar{X}_{\varrho 2}, \bar{X}_{\varrho 3}$, the body axis position coordinates (from the center of mass) to the center of pressure of the lifting surface,

2) d_ϱ, the X-axis Euler rotation angle of dihedral of the lifting surface, and

3) i_ϱ, the Y-axis Euler rotation angle of incidence of the lifting surface.

The lifting-surface-to-body-axis-coordinate transformation matrix can be defined by the two-axis Euler rotation matrix, $(c_\varrho) = X(d_\varrho)Y(i_\varrho)$, which becomes

$$(c_\varrho) = \begin{pmatrix} \cos i_\varrho & 0 & \sin i_\varrho \\ \sin i_\varrho \sin d_\varrho & \cos d_\varrho & -\cos i_\varrho \sin d_\varrho \\ -\sin i_\varrho \cos d_\varrho & \sin d_\varrho & \cos i_\varrho \cos d_\varrho \end{pmatrix} \tag{4.113}$$

The components of the undisturbed velocity of the air motion in station coordinates, as felt at the center of pressure of a wing lifting surface, are given by

$$v_{\varrho i} = b_i^\alpha \bar{W}_\alpha^\gamma \bar{X}_{\varrho\gamma} + v_i + v_{w_i} \tag{4.114}$$

where again, as given by Eq. (4.50), we have the $\bar{W}$ matrix, which is the skew-symmetric-body-axis-rotation-rate matrix given by

$$(\bar{W}) = \begin{pmatrix} 0 & -r & q \\ r & 0 & -p \\ -q & p & 0 \end{pmatrix} \tag{4.115}$$

Equation (4.114) is the same as Eq. (2.119) but includes the rotational velocity components of the lifting surface's center of pressure. The wing lifting surface relative airflow velocity components in body axis coordinates become

$$\bar{v}_{\varrho i} = \bar{W}_i^\gamma \bar{X}_{\varrho\gamma} + b_{\gamma i}(v^\gamma + v_w^\gamma) \tag{4.116}$$

The ϱth lifting surface relative air motion velocity components are

$$\bar{\bar{v}}_\varrho^i = (c_\varrho)_\alpha^i \bar{W}^{\alpha\gamma} \bar{X}_{\varrho\gamma} + (c_\varrho)_\alpha^i b_\gamma^\alpha (v^\gamma + v_w^\gamma) \tag{4.117}$$

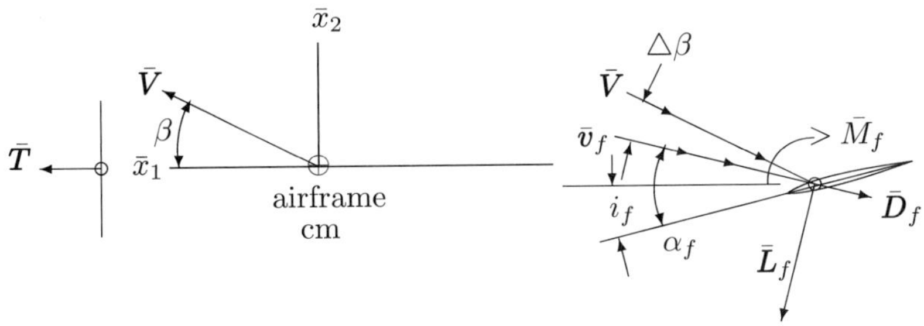

Fig. 4.14 Parameters defined by the geometry of the airplane vertical fin.

The notation is confused here in that ϱ is not a subscript but is used to identify the ϱth lifting surface and v_w^{γ} are the vector components of the atmospheric winds in station coordinates. Using these velocity components in Eqs. (2.120) and (2.121), we have the estimates for α_ϱ, β_ϱ, and $\bar{q}_\varrho$ at the lifting surface's center of pressure.

Looking down from the top, Fig. 4.14 represents, with exaggerated wind angles, the forces and moments on the airplane due to the vertical fin, defined by $d_\varrho =$ 90 deg. From this figure and as in Eq. (4.59), the angle of attack of the vertical fin is

$$\alpha_f = \beta - \Delta\beta - i_f \tag{4.118}$$

where β is the airplane sideslip angle given by Eq. (2.121), and $\Delta\beta$ is the sidewash angle due to propeller thrust, which causes a rotation of the slip stream behind the propeller and is a strong function of the engine power setting and the airplane forward speed, i.e., δ_t and $\bar{V}$. Note carefully the signs as shown here and that a positive angle of attack on the vertical fin will produce a lifting force in the negative $\bar{x}_2$ (airplane y axis) direction. The angle $\Delta\beta$ is considered positive when the propeller-induced airflow is in the positive $\bar{x}_2$ direction (Ref. 19, p. 294). The angle of incidence of the fin is i_f, and a negative angle is shown in Fig. 4.14. The fin angle of incidence is usually set such that the fin lift force will be zero at the airplane's cruising speed with a near-zero sideslip angle, i.e., with $\beta = 0$.

Part 2: Estimating the body axis aerodynamic forces including the effects of the control effectors. Using α_ϱ and β_ϱ as computed from Eq. (4.117), we can compute the ϱth wind-axis-to-lifting-surface transformation matrix, $(b_w)_\varrho$, as given by Eq. (2.122). Realizing that for the lifting surfaces in the air flowfields behind the airplane wing, the angles of attack become $\alpha_\varrho - \Delta\alpha$ for the tail surfaces and $\beta_\varrho - \Delta\beta$ for the vertical fin, respectively. Finally, using $\bar{q}_\varrho$ as previously computed, the lift force, drag force, and pitching moment magnitudes, given in Eq. (4.60), for each of the ϱth lifting surfaces can be converted into vector components in the lift, drag, and spanwise directions and transformed into their respective body axis vector components, given here in matrix form as follows:

$$\bar{\boldsymbol{L}}_\varrho = (c)_\varrho (b_w)_\varrho \begin{pmatrix} 0 \\ 0 \\ -\bar{L}_\varrho \end{pmatrix}$$

for the lift forces,

$$\bar{\boldsymbol{D}}_\varrho = (c)_\varrho (b_w)_\varrho \begin{pmatrix} -\bar{D}_\varrho \\ 0 \\ 0 \end{pmatrix}$$

for the drag forces, and the body axis thrust forces from Eqs. (4.69). The pitching moments from the ϱth surfaces become

$$\bar{\boldsymbol{M}}_\varrho = (c)_\varrho \begin{pmatrix} 0 \\ \bar{M}_\varrho \\ 0 \end{pmatrix} \tag{4.119}$$

Each of the body axis forces in Eqs. (4.119) must be transformed into the Earth-based station coordinates using Eq. (2.133). This transformation matrix is defined using the airplane's instantaneous yaw-pitch-roll Euler angles and is given in Appendix A.10. The contact forces in the Earth-based station coordinates can then be summed and placed in Eq. (2.114), forming the total acceleration vector components $\dot{\bar{v}}_i$ for the numerical solution for the translational motion of the airframe. These total acceleration components can be placed in the integration list in positions $\dot{y}_4$, $\dot{y}_5$, and $\dot{y}_6$, respectively, as shown in Appendix C.3, for the numeric solution process.

Part 3: Estimating the body axis aerodynamic moments acting on the airframe. The forces on the airframe are considered to act at the center-of-pressure points defined by each aerodynamic surface and the airplane's propulsion system. The static and dynamic stability requirements for the airplane constrain the locations of the force contact points relative to the center of mass of the airplane, or conversely, the airplane mass distribution and relative location of the force contact points become an integral part of the airplane design to guarantee the static and dynamic stability of the airplane. The contact points where airframe forces are considered to act are represented by the body axis position vectors from the center of mass, as previously defined. For this application, a simple database of airframe forces and *moment arms*, for n lifting surfaces, might be organized by using Table 4.5.

Table 4.5 Database of airframe forces and moment arms for n lifting surfaces

Airframe forces	Moment arms
1) Engine thrust force	$\bar{\boldsymbol{T}} : \bar{\boldsymbol{X}}_1$
2) Right wing lift and drag forces	$\bar{\boldsymbol{L}}_2$ and $\bar{\boldsymbol{D}}_2 : \bar{\boldsymbol{X}}_2$
3) Left wing lift and drag forces	$\bar{\boldsymbol{L}}_3$ and $\bar{\boldsymbol{D}}_3 : \bar{\boldsymbol{X}}_3$
4) Horizontal tail lift and drag forces	$\bar{\boldsymbol{L}}_4$ and $\bar{\boldsymbol{D}}_4 : \bar{\boldsymbol{X}}_4$
5) Vertical fin lift and drag forces	$\bar{\boldsymbol{L}}_5$ and $\bar{\boldsymbol{D}}_5 : \bar{\boldsymbol{X}}_5$
⋮	⋮
n) nth lifting surface lift and drag forces,	$\bar{\boldsymbol{L}}_n$ and $\bar{\boldsymbol{D}}_n : \bar{\boldsymbol{X}}_n$

Each force acting on the airframe body axes, as shown in Figs. 4.13 and 4.14, will cause a moment or a torque about the airframe center of mass. The total body axis moments can be described using the following vector equation:

$$\bar{N} = \sum_{\kappa=1}^{n} \bar{M}_{\kappa} + (\bar{X}_1 \times \bar{T}) + \sum_{\kappa=2}^{n} (\bar{X}_{\kappa} \times \bar{L}_{\kappa}) + \sum_{\kappa=2}^{n} (\bar{X}_{\kappa} \times \bar{D}_{\kappa}) \tag{4.120}$$

where $\bar{M}_{\kappa}$ is from the last equation of Eqs. (4.119).

Part 4: Computing the total body axis torques using Euler's equations of motion. The vector $\bar{N}$ has the components ($\bar{N}_1, \bar{N}_2, \bar{N}_3$), as shown in the second set of equations of Eq. (4.44). These moments become the driving function for the solution of Euler's equations of motion, given by Eq. (3.82), whose numeric solution will define the rotational dynamics for the airframe. The body axis rotation rate accelerations, the $\dot{\omega}_i$ components, are then placed into the integral list in positions $\dot{y}_{11}$, $\dot{y}_{12}$, and $\dot{y}_{13}$, respectively, as shown in Appendix C.3 for the numeric solution for the six-DOF airframe motion.

4.3 Flight Vehicle Motion Simulations

4.3.1 Cowell's Solution Method

The solution of the flight vehicle's equations of motion describing the accelerations of the center of mass of the vehicle, using stepwise numerical integration methods, is known as Cowell's method[59]. This method of propagation is generally used in three-DOF simulations where the rotational motion is not considered and does not effect the motion of the center of mass of the space vehicle. When using Cowell's method to propagate space vehicle motion, all of the perturbing accelerations can be included, and a very accurate simulation of the motion results. We can include, for instance, the perturbing effects of the nonspherical gravitational fields given by Eqs. (4.20), the effects of the atmospheric forces that may be acting on the vehicle as given by Eq. (4.106), and all other forces that we are able to describe in the differential equations of motion. Because many astrodynamic equations of motion have no closed-integral solutions, Cowell's method must be applied. Modern digital computer systems are now available and are designed for these types of problems. These computer systems make it possible to solve even the most complicated differential equations of motion using Cowell's method.

The integral lists given in Appendix C.1 can be used to describe the state vectors for each center of mass in the numerical solution of the differential equations of motion. These differential equations of motion should be written to be continuous over the step-size intervals. For instance, the simulation step size should be selected small enough so that control function event changes can be scheduled only at full-step intervals. Simple examples of the stepwise numerical solutions are given in Appendix D and can be applied to Cowell's method of solution.

The computer algorithms given in Appendix D are the simplest of the numerical methods and are called Cauchy-Euler methods, which require a minimum of functional evaluations of the equations of motion per time step. Higher order methods, requiring more evaluations of the equations of motion per step, such as Runge-Kutta methods, are described in Ref. 8 (pp. 500–502). The Runge-Kutta methods are used extensively for solutions of three-DOF motion, as well as six-DOF simulations that include the solution for the rotational motion of aerospace vehicles.

By the late 1960s the predictor-corrector methods of Adams-Moulton and Gauss-Jackson were being applied to the three-DOF motion of orbiting space vehicles using large-scale digital computer systems. These methods are outlined in Ref. 29 (Sec. 6.6) and Ref. 8 (p. 505), respectively. Large time steps on the trajectory are used to create a span of back-order solution points that are then used to fit an accurate solution for the next projected time step. These solution methods require the differential equations of motion to be continuous for longer periods of time and hence *are not generally applied to six-DOF motion.*

4.3.2 Two-Body Orbital Motion Solution Method

The two-body orbital motion problem addresses the use of an approximating function for the perturbed gravitational potential, such as in Eq. (4.7), instead of, for example, the gravitational potential function from n-body rotating mass points. Recent work by Bond and Gottlieb[48] is the most outstanding and perhaps the finest analysis for the solution of the two-body orbital equations of motion. At the time of this publication, this work represents the highest level of achievement in the historical development for the numerical solution of the two-body orbital motion problem. A detailed discussion of the Bond and Gottlieb method is beyond the scope of this text; however, some of the basic mathematical concepts used in the analysis are presented as follows.

The perturbations to the motion of the space vehicle due to the effects of the rotating gravitational field of the planet are based on the assumption that the gravitational potential function will be equal in both the inertial and the rotating coordinate frames (Ref. 27, pp. 139, 140). Starting with Eq. (4.6) in functional form, we have

$$\mathcal{V} \approx \mathcal{V}(x_\alpha, t) = \mathcal{V}(\bar{x}_\alpha) \tag{4.121}$$

Differentiating this equation with respect to the independent variable time, we have

$$\frac{\mathrm{d}\mathcal{V}}{\mathrm{d}t} = \frac{\partial \mathcal{V}}{\partial x^\alpha}\dot{x}_\alpha + \frac{\partial \mathcal{V}}{\partial t} = \frac{\partial \mathcal{V}}{\partial \bar{x}^\alpha}\dot{\bar{x}}_\alpha \tag{4.122}$$

This is the same analysis procedure used to derive Eq. (4.16). In vector form Eq. (4.122) is written as

$$\frac{\mathrm{d}\mathcal{V}}{\mathrm{d}t} = \nabla\mathcal{V}\cdot \boldsymbol{v} + \frac{\partial \mathcal{V}}{\partial t} = \nabla\mathcal{V}\cdot \boldsymbol{\nu} \tag{4.123}$$

Using Eq. (1.228), the planet relative velocity components in the inertial frame, namely, $\nu^\rho = W^\rho_\alpha x^\alpha + \dot{x}^\rho$, can be written in vector form as

$$\boldsymbol{\nu} = \boldsymbol{\omega} \times \boldsymbol{r} + \boldsymbol{v} \tag{4.124}$$

where $\boldsymbol{\omega}$ is the rotation rate vector of the planet. In the case of the Earth, $\boldsymbol{\omega} = (0, 0, \omega_e)$, where ω_e is the sidereal rotation rate of the Earth as discussed in Sec. 2.3.2 of the text. Placing Eq. (4.24) into Eq. (4.123) and realizing that the partials with respect to position will cancel one another, we have the interesting relationship for the rate of change of the potential function with respect to time:

$$\frac{\partial \mathcal{V}(\boldsymbol{r}, t)}{\partial t} = -\boldsymbol{\omega} \times \boldsymbol{r} \cdot \left(\frac{\partial \mathcal{V}(\boldsymbol{r}, t)}{\partial \boldsymbol{r}} \right) = -\boldsymbol{\omega} \cdot \boldsymbol{r} \times \left(\frac{\partial \mathcal{V}(\boldsymbol{r}, t)}{\partial \boldsymbol{r}} \right) \tag{4.125}$$

Using the total energy relationship (per unit mass) and its derivative with respect to time,

$$\frac{\mathrm{d}}{\mathrm{d}t}(1/2v^2 + \mathcal{V}) = \frac{\partial \mathcal{V}}{\partial t} \tag{4.126}$$

and Eq. (4.125), we have

$$1/2v^2 + \mathcal{V}(\boldsymbol{r}, t) - \boldsymbol{\omega} \cdot (\boldsymbol{r} \times \boldsymbol{v}) = \text{const} \tag{4.127}$$

Notice that $(\boldsymbol{r} \times \boldsymbol{v})$ is the angular momentum vector of the orbit as discussed in Eq. (2.50). This constant, as given in Eq. (4.127), is named the *Jacobi integral* because of its similarity to the Jacobi integral used in the solution of the *restricted three-body problem* as described in Ref. 60.

The Sundmann transformation Ref. 61 is used in the analysis to transform the independent variable time t to the fictitious time variable s, with the defining relationship

$$\frac{\mathrm{d}s}{\mathrm{d}t} = \frac{1}{r} \tag{4.128}$$

where, $r = |\boldsymbol{r}|$. Using this fictitious time variable s, the components of the inertial velocity vector are transformed as follows:

$$v_i = \frac{\mathrm{d}r_i}{\mathrm{d}s}\left(\frac{\mathrm{d}s}{\mathrm{d}t}\right) = \frac{\mathrm{d}r_i}{\mathrm{d}s}\left(\frac{1}{r}\right) \tag{4.129}$$

Similarly, the acceleration vector components are transformed into a function of s and become

$$\frac{\mathrm{d}^2 r_i}{\mathrm{d}t^2} = \frac{\mathrm{d}}{\mathrm{d}s}\left(\frac{1}{r}\frac{\mathrm{d}r_i}{\mathrm{d}s}\right)\frac{\mathrm{d}s}{\mathrm{d}t} = \left(\frac{1}{r^2}\right)\frac{\mathrm{d}^2 r_i}{\mathrm{d}s^2} - \left(\frac{1}{r^3}\right)\left|\frac{\mathrm{d}\boldsymbol{r}}{\mathrm{d}s}\right|\frac{\mathrm{d}r_i}{\mathrm{d}s} \tag{4.130}$$

The Bond and Gottlieb formulation uses the Jacobi integral, which is embedded in the Newtonian equations of motion. The Jacobi integral is a constant for the case when the two-body system is perturbed by a potential function that is explictly dependent on time as well as position. The method develops 14 independent equations Ref. 75 as functions of the fictitious time variable s, and hence is named BG14. The equations are integrated simultaneously to a new fictitious time s, and transformed back into the inertial position and velocity components at the new time t.

Using the BG14 solution methods, time steps on the order of fractions of a day can be taken even when simulating space vehicle motion in low Earth orbit. Application of the BG14 methods provides the most accurate solutions to the two-body orbital equations of motion and also makes very efficient use of computing resources.

4.3.3 Aerospace Applications: Six-DOF Motion Simulation

The analysis methods presented in previous sections of the text make it possible to describe the solution of typical six-DOF motion problems using computer simulations. The following analysis can be used to simulate the motion of flight vehicles and their onboard GNC systems for engineering analysis and testing. Using modular software concepts, this analysis can be adapted to a wide variety of aerospace motion simulations. The following flight regimes, requiring specific flight environments and different GNC disciplines, can be identified for analysis by the aerospace engineer: 1) atmospheric flight; 2) rocket-powered ascent from the launch pad to orbit injection; 3) on-orbit operations, rendezvous, and docking; 4) interplanetary trajectory motion; and 5) de-orbit and atmospheric entry flight.

The logic flow diagram presented in Fig. 4.15 shows the basic order and defines some of the functional blocks used to simulate the six-DOF motion of an arbitrary flight vehicle. We will call this example *flight simulation* by the acronym FLTSIM. In block 2, the word *Segment* refers to the specific flight period and the differential equations of motion that can occur sequentially in the same run. The *Segment Loop* is performed for each flight condition being studied. For example, a contiguous flight simulation might be desired that would include an aircraft takeoff roll, climbing flight, and continuing to the cruise condition. Another typical application is a rocket ascent trajectory with rocket staging to eventual orbit injection. Each flight segment or mode of flight requires different differential equations of motion to be integrated.

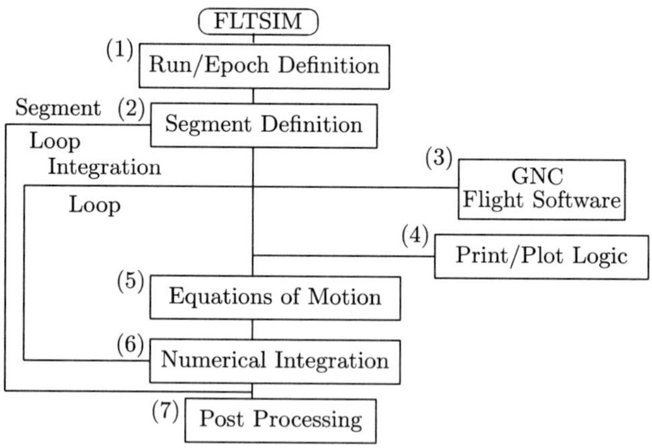

Fig. 4.15 Six-DOF flight simulation logical flow diagram.

The *Integration Loop* is cycled at the step size Δt as defined by the numerical integration process being applied. A single time step is performed in this loop, i.e., the solution of the differential equations of motion is advanced from t_n to $t_{n+1} = t_n + \Delta t$. Depending on the motion being studied, simulated motion of the vehicle's center of mass can be achieved with time steps in the order of seconds. However, simulation of the GNC functions may require evaluations at a much higher frequency, for example, 25 Hz (cycles per second) for sensitive flight control systems.

Based on Fig. 4.15, each of the functional blocks can be defined to perform the following:

1) Run/Epoch Definition:

a) Base epoch and the vehicle initial state are established; position and velocity vectors are defined;

b) Vehicle body axis attitude and rotation rates are determined; and

c) GNC parameters are initialized, such as sensor positions and the REFMAT and IMU gimbal angles.

2) Segment Definition:

a) Integration control parameters are input, such as segment start and stop times;

b) The differential equations of motion to be integrated during this flight period are defined;

c) Integrator is selected with base cycle time;

d) Gravity and atmospheric models are selected;

e) Vehicle weight and body-axis-moment-of-inertia matrix are defined; and

f) GNC cycle times and control parameters are determined.

3) GNC Flight Software (refer to Sec. 1.4.8):

a) Sensor simulation determines estimated measurement data;

b) Actual navigation software computes the navigated state parameters to be used by the guidance and flight control systems; and

c) Guidance and flight control functions compute the movement of the vehicle's *control effectors* to fly the aerospace vehicle to the desired trajectory. For example, the control effectors could be the aerodynamic control surfaces, rocket engine gimbal angles, or rocket engine on-off firing times. The control effectors are moved to cause the forces and moments to act on the vehicle to fly the desired trajectory and flight attitude.

4) Print/Plot Logic:

a) Data files are written as the segment integral process continues, containing the print and plot parameters intended for the Post Processing block for the presentation of the flight performance information.

5) Equations of Motion:

a) The differential equations of motion selected for this flight segment are used to compute the accelerations of the vehicle's center of mass and the rate of change of the body axis rotation rates (refer to the integral list given in Appendix C.3);

b) This block can call the gravity model as given in Eqs. (4.20), the atmospheric model to determine the flight density to compute the drag force given in Eq. (4.106), and the rocket engine thrust effects given in Eq. (4.82); and

c) The rate of change of the body axis rotation rates are given by Euler's rotational equations of motion as discussed in Sec. 3.2.5 and as given in Eqs. (3.42).

6) Numerical Integration:

a) The vehicle differential equations of motion are numerically integrated in one time step; and

b) The integral loop will continue to cycle until the segment stop time or limiting boundary condition is reached.

7) Post Processing:

a) The print and plot files are sorted, and the resulting performance parameters are printed and graphically presented for analysis by the aerospace engineer.

4.4 Space Vehicle Motion Using Mean Orbital Elements

4.4.1 Perturbation Analysis Methods and Mean Orbital Elements

Perturbation methods are a study topic in the solution of differential equations and have many applications in the analysis of science and engineering problems. In this application, it is the detailed analysis of the variant characteristics of the Keplerian orbital elements. In Sec. 2.2.3, we showed that the Keplerian orbital elements can be solved using the differential equations of motion for a space vehicle in orbit about a planet with a central gravity field. Equation (2.48) is repeated here for continuity in the discussion and is written as follows:

$$\ddot{x}_i + \frac{\mu}{r^3} x_i = 0 \tag{4.131}$$

The Keplerian orbital elements are represented in functional notation at the defining epoch by

$$e_i = e_i(\boldsymbol{r}_o, \boldsymbol{v}_o, t_o) \tag{4.132}$$

Likewise, in Sec. 2.2.4, we showed that the space vehicle position and velocity states can be propagated in its orbit to a new epoch time t by

$$x_i(t) = x_i(e_\alpha(t_o), t) \tag{4.133}$$

and

$$\dot{x}_i(t) = \dot{x}_i(e_\alpha(t_o), t) \tag{4.134}$$

respectively.

When simulating the orbital motion of a space vehicle using Cowell's method, we can include the perturbing accelerations due to the nonspherical gravity harmonics and other accelerations from forces such as aerodynamic drag. We can write the equations of motion for the space vehicle as follows:

$$\ddot{x}_i = -\left(\frac{\mu}{r^3}\right) x_i + \nabla \mathcal{R}_i + \mathcal{P}_i \tag{4.135}$$

where the $\mathcal{P}_i$ are the components of all of the other perturbing accelerations combined. Unlike the solution for the Keplerian orbital elements, as described in

Sec. 2.2.4 and by Eqs. (4.133) and (4.134), there is no *closed-form* solution for Eq. (4.135). Numerical methods must be used, which can require large amounts of computing resources, especially when long-term integrals of the motion are necessary for the solution of aerospace problems. One such problem is the long-term aerodynamic drag effects on space vehicle orbits, for example, the slow decay of the orbits of orbiting space stations.

Notice that in Keplerian motion, the orbital elements are constant as the space vehicle moves along its orbital path. One of the basic concepts of this perturbation analysis application is that the Keplerian elements can be solved for along the perturbed orbit defined by Eq. (4.135). That is, at each point in time, the *osculating orbit* can be re-defined by the Keplerian solution of Eq. (4.132). When this procedure is done, the Keplerian orbital elements along the real orbit defined by Eq. (4.135) are no longer constant and osculate around some set of *mean orbital elements*.

For the classic example, using only the nonspherical gravity perturbations, we can differentiate Eq. (4.134), and by substitution Eq. (4.135) becomes

$$\frac{\partial \dot{x}_i}{\partial e_\alpha}\dot{e}^\alpha + \frac{\partial \dot{x}_i}{\partial t} + \left(\frac{\mu}{r^3}\right) x_i = \nabla \mathcal{R}_i \tag{4.136}$$

This equation forms three component equations. By differentiating Eq. (4.133), we can write

$$\frac{\mathrm{d}x_i}{\mathrm{d}t} = \frac{\partial x_i}{\partial e_\alpha}\dot{e}^\alpha + \frac{\partial x_i}{\partial t} \tag{4.137}$$

Realizing that along the perturbed orbit $\dot{e}_\alpha \neq 0.0$, it may be possible to define the case where

$$\frac{\partial x_i}{\partial e_\alpha}\dot{e}^\alpha = 0 \tag{4.138}$$

From Eq. (4.137) and from differentiating Eq. (4.134), we see that along the Keplerian orbit we have

$$\frac{\mathrm{d}x_i}{\mathrm{d}t} = \frac{\partial x_i}{\partial t} \tag{4.139}$$

and

$$\frac{\mathrm{d}\dot{x}_i}{\mathrm{d}t} = \frac{\partial \dot{x}_i}{\partial t} \tag{4.140}$$

Along the instantaneous osculating orbit, which we define as crossing the true orbit, we have

$$\frac{\partial \dot{x}_i}{\partial t} + \left(\frac{\mu}{r^3}\right) x_i = 0 \tag{4.141}$$

which is true from Eq. (4.131). Using Eq. (4.141) in Eq. (4.136), we have

$$\frac{\partial \dot{x}_i}{\partial e_\alpha}\dot{e}^\alpha = \nabla \mathcal{R}_i \tag{4.142}$$

Equations (4.138) and (4.142) define six simultaneous differential equations with six unknowns, namely the $\dot{e}_\alpha$. Solution of these equations was done by Lagrange in 1774. This very elegant solution is known as *Lagrange's planetary equations* and is summarized here as follows:

$$
\begin{aligned}
\dot{a} &= \left(\frac{2}{na}\right)\frac{\partial \mathcal{R}}{\partial M}\\
\dot{e} &= \left(\frac{1-e^2}{na^2 e}\right)\frac{\partial \mathcal{R}}{\partial M} - \left(\frac{\sqrt{1-e^2}}{na^2 e}\right)\frac{\partial \mathcal{R}}{\partial \omega}\\
\dot{i} &= \left(\frac{\cos i}{na^2 \sin i\sqrt{1-e^2}}\right)\frac{\partial \mathcal{R}}{\partial \omega}\\
\dot{\Omega} &= \left(\frac{1}{na^2 \sin i\sqrt{1-e^2}}\right)\frac{\partial \mathcal{R}}{\partial i}\\
\dot{\omega} &= -\left(\frac{\cos i}{na^2 \sin i\sqrt{1-e^2}}\right)\frac{\partial \mathcal{R}}{\partial i} + \left(\frac{\sqrt{1-e^2}}{na^2 e}\right)\frac{\partial \mathcal{R}}{\partial e}\\
\dot{M} &= n - \left(\frac{1-e^2}{na^2 e}\right)\frac{\partial \mathcal{R}}{\partial e} - \left(\frac{2}{na}\right)\frac{\partial \mathcal{R}}{\partial a}
\end{aligned}
\tag{4.143}
$$

See Refs. 62 and 63 for more detail in the derivations of Eqs. (4.143).

4.4.2 Brief Summary of Propagation Methods Using Mean Orbital Elements

The Cartesian states of orbiting space vehicles change vary rapidly as the vehicle moves along its orbit. Numerical integration using Cowell's solution methods is necessary to accurately predict the motion, especially in low Earth orbits. On the other hand, the orbital elements change slowly and display definite cyclic characteristics. The early analysis work of Kozai [64] showed that mean orbital element sets could be derived that include perturbation effects of the first three or four zonal harmonics. These first zonal harmonics cause the largest deviations from the Keplerian orbit. Since that time, perturbation analysis procedures have continued to be developed and are now used by the aerospace engineer and analyst to approximate the motion of space vehicles. Works by Ward[65] in the late 1960s applied the methods of Kozai to predict the motion of Earth-orbiting satellites accurately enough to schedule tracking station antennas for acquisition and tracking. Continued development of these solution methods, employing greater amounts of tracking data, led to the *NORAD element set*,[66] which can predict the position of the Earth-orbiting space vehicle within several kilometers. However, this accuracy can only be achieved relatively close to the defining epoch of the mean element set, say within several days of the defining epoch. Other perturbations, such as those caused by aerodynamic drag and the higher order gravity harmonics, cause the predicted motion to deviate from the real motion of the space vehicle.

Operational use of the mean orbital elements for motion prediction is used when it is necessary to predict the position of many orbiting space vehicles. Using the notation $\bar{e}(t_o)$ for the mean orbital element set, we can describe their application with the following simplified analysis concepts. The Keplerian orbital elements, at time t, can be estimated from the mean element set, which represents the orbit by the functional equation:

$$e_i(t) = \bar{e}(t_o) + \delta e_{\text{sec}} + \delta e_{\text{sp}} + \delta e_{\text{lp}} \tag{4.144}$$

From the analysis of the nature of the derivatives of the mean orbital elements, i.e., the $\dot{e}_\alpha$ from the previous section, it can be shown that three distinct types of variations can be made to represent the osculating orbital elements. In Eq. (4.144) we have the following:

1) the δe_{sec}, which are called the secular variations, the variations that are functions of time since epoch;

2) the δe_{sp}, which are the short-period variations, those variations that are cyclic within a single orbit; and

3) the δe_{lp}, which are the long-period variations, those variations that are cyclic over many orbit cycles.

These variations are derived as functions of the zonal harmonics and the orbital elements themselves. We complete the process using the Keplerian elements from Eq. (4.144), and the point transformation is discussed in detail in Sec. 2.2.4, which can be made at any time t along the orbit.

The analysis works of Refs. 67, 68, 69, and 70 *to give due credit to a few*, develop methods of calculating the perturbing effects to the motion using the concepts of the mean orbital element set. Frazer[71] developed methods of calculating the perturbations in Cartesian coordinates. Frazer's method does not have the problems with near-circular orbits as does the Kozai methods because the variations are computed in Cartesian form. It is also noteworthy that Frazer's computational algorithms (Ref. 47, Sec. 3.0) closely match the short- and long-period osculations when compared to the numerically integrated results using Cowell's method.

The secular rates of change of the mean orbital elements, i.e., δe_{sec} terms in Eq. (4.144), are briefly summarized here as follows:

1) $\dot{a}$, the orbit semimajor axis, from Ref. 26 (p. 64),

$$\frac{\dot{a}}{a} = \left(\frac{2B\bar{q}g_o}{na}\right)\left(\frac{\sqrt{1+e^2+2e\cos\nu}}{\sqrt{1-e^2}}\right) \tag{4.145}$$

The change in the orbit mean semimajor axis represents the change in orbital energy. Notice that as the orbit eccentricity e approaches zero in Eq. (4.145), we have Eq. (4.112), which was derived differently using the orbit energy relationships for circular orbits. The perturbation due to aerodynamic drag removes energy from the orbit, thus causing a slow decrease in the orbit semimajor axis. When the orbit semimajor axis decays and falls below about 1.03 Earth radii, reentry into the Earth's atmosphere is imminent, and propagation using perturbation analysis is no longer valid. Notice that if the ballistic coefficient B is zero, the mean semimajor axis will remain unchanged over time, but will osculate in value during the orbit due to the long- and short-period variations.

2) $\dot{e}$, the orbit eccentricity, also from Ref. 26,

$$\dot{e} = \left(\frac{-2B\bar{q}g_o}{na}\right)\left(\frac{\sqrt{1-e^2}(\cos\nu + e)}{\sqrt{1+e^2+2e\cos\nu}}\right) \tag{4.146}$$

The space vehicle's orbital eccentricity is slowly decreased by aerodynamic drag, causing the orbit to become more and more circular in time.

3) $\dot{i}$, the orbit inclination; the secular variation in orbital inclination is very small and for most Earth-orbiting applications, the orbital inclination can be considered constant.

4) $\dot{\Omega}$, the right ascension of the orbit ascending node, from Ref. 64 (p. 372),

$$\dot{\Omega} = -\frac{3}{2}\left(\frac{J_2\cos i}{p^2}\right)n \tag{4.147}$$

where $p = (a/a_e)(1-e^2)$ is the parameter of the orbit in Earth radii. The J_2 harmonic is the predominant perturbation causing the orbital plane to precess in the inertial coordinate frame. It is interesting to point out here that the aerospace engineer uses Eq. (4.147) in the orbit mission design for sun-synchronous orbits. The sun's apparent motion along the ecliptic plane is about 0.986 deg per day. By placing the sun's apparent rate in the left side of Eq. (4.147), a family of sun-synchronous orbits is defined by the relationship between the orbit semimajor axis a and the orbital inclination i. These orbits must be slightly retrograde, i.e., $i > 90$ deg. For an excellent discussion for the design of sun-synchronous orbits, see Ref. 8 (Sec. 11.4.1).

5) $\dot{\omega}$, the argument of perigee, also from Ref. 64;

$$\dot{\omega} = \frac{3}{2}\left(\frac{J_2}{p^2}\right)\left(2 - \frac{5}{2}\sin^2 i\right)n \tag{4.148}$$

Notice that at the critical inclination of 63.435 deg, the factor $(2 - (5/2)\sin^2 i)$ will be zero and become negative for more highly inclined orbits.

6) n, Kozai's mean motion, also from Ref. 64,

$$n = n_o\left[1 + \frac{3}{2}\left(\frac{J_2}{p^2}\right)\left(1.0 - \frac{3}{2}\sin^2 i\right)\sqrt{1-e^2}\right] \tag{4.149}$$

where $n_o = \sqrt{\mu}/a^{(3/2)}$, the Keplerian mean motion.

4.4.3 Aerospace Applications: Estimating Close Encounters Between Orbiting Space Objects

Since the early 1960s, there have been many thousands of space vehicles launched into Earth orbit. Many of these space vehicles, whether in use or fallen into disuse, are still in orbit today and are likely to remain in orbit for many years to come. However, as a space object's orbit slowly decays due to aerodynamic drag forces, many objects are purged from Earth orbit and reenter the atmosphere and simply burn up. There is a possibility that parts of the space vehicle could survive reentry and contact the Earth's surface.

Using the methods of orbit determination, a number of databases of mean orbital elements have been established representing the orbits of many space vehicles and objects that remain in orbit around the Earth. The U.S. Space Command database contains mean element sets for the orbits of over 8,000 space objects (at the time of this writing). Generally, the space objects that are tracked and form the database are objects greater than 10 cm in size. With so many space objects in Earth orbit and with the length of time an active space vehicle, such as a space station, might be in orbit, close encounters or even collisions are possible.

One of the collision prediction methods used today is to estimate the positions of space objects in their orbits using mean orbital elements. The mean orbital element sets can be determined using minimum variance least-squares solutions that fit the space object's orbit to actual tracking data. Using the same propagation methods that were used by the orbit determination processor, the orbits of the space objects can be propagated forward in time from their solution epochs. From the discussions in Sec. 4.4.2, the Keplerian (osculating) orbital elements for the space object's position and velocity solutions can be estimated at time t from the mean orbital elements $\bar{e}_i$, which have been determined at the epoch time t_o. The analytic solution process is represented by Eq. (4.144).

From the geometry of the inertial orbit paths, the procedures[72] in the following sections can be used to predict which space objects can be *in conjunction with* or create close encounters with the space vehicle of interest. These sections describe an operational process that uses the U.S. Space Command (NORAD) database. The orbits of the low-altitude space objects are propagated using the SGP4 solutions[66] and the SDP4 solutions[66] for the high-altitude objects with periods of greater than 225 min. This collision prediction process is best described as a series of five editing procedures, and after each edit the remaining space objects are only those that could create close encounters with the space vehicle of interest. We will refer to the space vehicle of interest as the *space vehicle* (SV) and the *space objects* (SO) as those being tested for possible conjunction situations. The following sections outline a conceptual description of the editing process, omitting the bookkeeping details required for the operational program system.

4.4.3.1 Perigee-apogee edits. The mean orbital elements of both the SV and the SO are propagated to the epoch time of interest to begin the analysis. Only the secular perturbed orbital elements are used here, since they better describe the mean orbits in space. Using Eq. (2.69), the orbit radii at perigee ($\nu = 0$) and apogee ($\nu = \pi$), respectively, are given by

$$r_P = a(1 - e) \tag{4.150}$$

and

$$r_A = a(1 + e) \tag{4.151}$$

Space objects with perigee radii (here, some tolerance is added to account for the short- and long-period osculations in the orbit radius) that are greater than the apogee radius of the space vehicle cannot create close encounter situations. Likewise, space objects with apogee radii less than the perigee radius of the space vehicle can be eliminated from the subset of threatening space objects.

4.4.3.2 Radius edits at the orbital plane intersection. Using the secular perturbed orbital elements, as in Sec. 4.4.3.1, and the orbit longitude coordinate transformation to inertial coordinates, the orbital inclination vector in the inertial frame is given by

$$\boldsymbol{i} = Z(\Omega)X(i)\begin{pmatrix}0\\0\\1\end{pmatrix} = \begin{pmatrix}\sin\Omega\sin i\\-\cos\Omega\sin i\\\cos i\end{pmatrix} \tag{4.152}$$

The orbital planes of the SV and an SO intersect along the lines defined by

$$\mathcal{N} = \pm(\boldsymbol{i}_{\mathrm{SV}} \times \boldsymbol{i}_{\mathrm{SO}}) \tag{4.153}$$

The Laplace vector or the vector to the orbit's perigee, as discussed in Sec. 2.2.3 and in Eq. (2.65), is defined using the orbital-coordinates-to-inertial-coordinates transformation as follows:

$$\mathcal{P} = Z(\Omega)X(i)Z(\omega)\begin{pmatrix}1\\0\\0\end{pmatrix} \tag{4.154}$$

The angles between the intersection line vectors, $\mathcal{N}$, and the Laplace vectors of the space vehicle, $\mathcal{P}_{\mathrm{SV}}$, and the space objects, $\mathcal{P}_{\mathrm{SO}}$, respectively, define four true anomalies to the intersection lines of the two orbital planes. Using these true anomalies in Eq. (2.69), we can estimate the orbit radii of the SV and that of the SO at the orbit intersection lines. The difference in the orbit radii along the orbit plane intersection lines is given by

$$\Delta r_k = |r_{\mathrm{SV}_k} - r_{\mathrm{SO}_k}|, \qquad \text{where } k = 1, 2 \tag{4.155}$$

Here the subscript k depicts each of the two orbital intersection lines. If the Δr_k exceed a defined tolerance, this orbit intersection created by the SO cannot threaten the SV and can be omitted from further analysis.

Notice from Eq. (4.154) that if the orbits of the SV and SO are near coplanar, special analysis is required. This coplanar analysis is not included here due to the length of the discussion.

The estimated times of arrival at the orbit intersection lines for both the SV and the SO can also be computed from the true anomalies just determined. The relationship between the true anomaly and the eccentric anomaly, as given in Eq. (2.76), is given by

$$\tan E = \frac{\sqrt{1-e^2}\,\sin\nu}{\cos\nu + e} \tag{4.156}$$

The mean anomalies $M_{t_{AR}}$ follow from Kepler's equation given in Eq. (2.80). Using the mean motion n, from Eq. (2.78), in Eq. (2.83), the estimated time of arrival at the orbit intersection line becomes

$$t_{AR} = t_o + \frac{M_{t_{AR}} - M_{t_o}}{n} \tag{4.157}$$

where t_o is the epoch time of the mean orbital element set, M_{t_o} is the mean anomaly of the element set at epoch, and n is the mean motion given approximately by Eq. (4.199).

4.4.3.3 Delta altitude edits. The SV orbit is propagated to the intersection arrival time, t_{SV_o}, and using the Cartesian position and velocity states, we compute the LVLH-to-inertial-coordinate transformation matrix, as described in Sec. 1.1.4. The SO is propagated to t_{SO_o}, and its inertial Cartesian position and velocity states are transformed to the SV's LVLH relative state with coordinates $(\bar{x}, \bar{y}, \bar{z}, \dot{\bar{x}}, \dot{\bar{y}}, \dot{\bar{z}})$. The relative motion trajectories of SOs in the SV LVLH coordinates are near-straight lines with $\dot{\bar{z}} \approx 0$. This is not true for meteoric particles that can approach the SV from any angle. Meteoric particles are coming from hyperbolic orbits that can have much higher flight path angles. The relative delta altitude ΔH can then be estimated as depicted in Fig. 4.16.

Notice that the distance, ΔH, is measured in the LVLH relative vertical direction. If ΔH at the closest approach point, as shown in Fig. 4.16, is greater than a defined tolerance, the conjunction situation does not threaten the SV with a possible collision.

4.4.3.4 Time of conjunction edits. Associated with each mean element set is an estimate of the errors in position at the epoch time and estimates for the growth of these errors as the mean elements are propagated from its epoch time. These errors can be visualized as an ellipsoid of revolution whose dimensions represent the down-range, cross-range, and altitude errors in position. The orbit

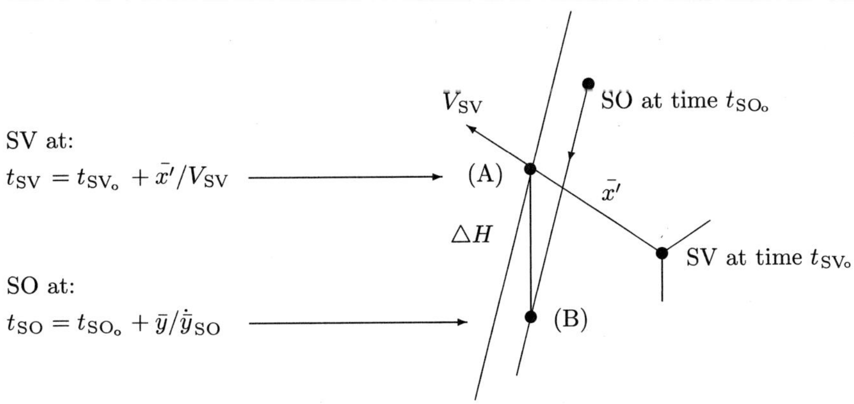

Fig. 4.16 The trajectory of the space object in the LVLH coordinates of the space vehicle.

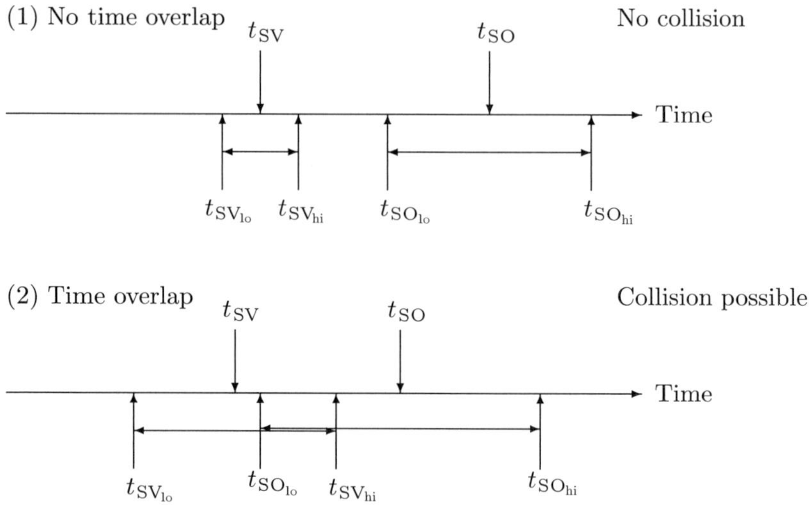

Fig. 4.17 The space vehicle and space object arrival times at the closest approach point.

intersection arrival times, t_{SV} and t_{SO}, for the SV and the SO, respectively, can be thought of as when the center of the error ellipsoid arrives at the closest approach points (A) and (B), as shown in Fig. 4.16. The down-range position error can be transformed into an error in time and added to adjust the estimated times of arrival, $t'_{SV} = t_{SV} \pm \Delta t_{SV}$ and $t'_{SO} = t_{SO} \pm \Delta t_{SO}$. For example, a down-range error of 4 n miles would give a Δt value of about 1 second in time for a low Earth orbit.

The error ellipsoid of the SV will reach the closest approach point (A), in Fig. 4.16, at time $t_{SV_{lo}} = t_{SV} - \Delta t_{SV}$ and exit this point at $t_{SV_{hi}} = t_{SV} + \Delta t_{SV}$. Likewise, the error ellipsoid of the SO can arrive at point (B) at $t_{SO_{lo}} = t_{SO} - \Delta t_{SO}$ and exit at $t_{SO_{hi}} = t_{SO} + \Delta t_{SO}$. Using Fig. 4.17, we can visualize a horizontal time scale, with time increasing from left to right on the page. Two conditions can be analyzed showing if the SV and the SO can occupy the orbit intersection at the same time. When there is no time overlap, as shown in condition 1, the SO can be omitted from further collision analysis. If there is a time overlap at the closest approach point, as in condition 2, the SV and SO could be in a collision situation.

4.4.3.5 Error ellipsoid edits. The final and most severe test for collision is the computation of the collision probability. If the error ellipsoid volumes, as described in Sec. 4.4.3.4, do not intersect each other, i.e., do not share a common volume, then the collision probability is zero. On the other hand, if the error ellipsoid volumes penetrate each other as they pass the orbital intersection point, a collision can occur and a collision probability can be estimated.

Because the position errors in the propagation methods used in this procedure can be large, actual collisions cannot be predicted. However, collision probabilities can be estimated in close encounter situations with orbiting space objects and are used to assess the threat to the space vehicle. If the collision probability estimate exceeds a defined threshold, it may be deemed necessary to change the orbit of an active space vehicle to avoid this conjunction situation altogether.

Conclusion

The gravitational accelerations acting on a flight vehicle caused by a planetary body can be modeled very accurately using the Legendre polynomial functions. However the gravity harmonic coefficients for the Legendre polynomials are not that easy to determined and must be solved for beforehand by using tracking data measurements from a known space vehicle in orbit around the planet. This method to model real world gravitational environments is applied in many aerospace vehicle motion simulations.

The Earth's atmosphere can be modeled accurately to an altitude of about 86 kilometers by using the perfect gas law and the hydrostatic equation for both temperature and density. Above 86 kilometers however, the total atmospheric molecular weight varies, and density and temperature become functions of a number of parameters, such as solar and geomagnetic activity. In these high altitude regions of the atmosphere, dynamic atmosphere models, such as that offered by L. G. Jacchia, must be used to estimate the temperature and density environments.

I have introduced the reader to some of the fundamentals for computing the fluid dynamic forces and moments on the flight vehicle using finite element modeling techniques. These methods began to be develop after 1965 as the aerodynamist had access to more computing power.

I have outlined the use of Cowell's solution methods and the techniques for the numerical solution of the six-DOF airframe motion applied in actual GNC flight simulators. This simulation method can use many of the acceleration, force, and moment models presented in this chapter to estimate the solution of the equations of motion of a flight vehicle.

I have also presented solution methods for the Two-body orbital motion problem that apply the methods of perturbation analysis and the use of the mean orbital elements in space vehicle motion analysis. In conclusion, I have offered the reader a discussion of the computational analysis supporting a ground-based collision avoidance software system. This example method was developed using the mean orbital elements to estimate collision situations between an orbiting space vehicle and the great number of Earth orbiting debris objects.

Appendix A: Relationships for Three-Axis Euler Rotational Sequences

Early in the aerospace program, especially as computer system resources became available to the aerospace engineer, Kimball[73] realized the importance of the 12 Euler rotation transformation matrices, which have now become an integral part of the analysis of the dynamics of flight vehicles. Today, the Euler transformation matrices remain an essential part of these analyses and are required in many aerospace applications. The 12 Euler transformation matrices, as described in Sec. 1.3.3, are presented here for each rotation sequence based on the single-axis rotation Eqs. (1.81), (1.82), and (1.83). The Euler matrices, as presented in this appendix, transform vectors from the system that has been rotated into vectors in the system considered to be stationary. Also presented are the equations for the quaternion as a function of the Euler angles and the Euler angles as a function of the matrix elements for each rotation sequence.[74]

A.1 Axis Rotation Sequence: 1, 2, 3

$$M = M(X(\theta_1), Y(\theta_2), Z(\theta_3)) = XYZ$$

$$M = \begin{pmatrix} \cos\theta_2\cos\theta_3 & -\cos\theta_2\sin\theta_3 & \sin\theta_2 \\ \sin\theta_1\sin\theta_2\cos\theta_3 + \cos\theta_1\sin\theta_3 & -\sin\theta_1\sin\theta_2\sin\theta_3 + \cos\theta_1\cos\theta_3 & -\sin\theta_1\cos\theta_2 \\ -\cos\theta_1\sin\theta_2\cos\theta_3 + \sin\theta_1\sin\theta_3 & \cos\theta_1\sin\theta_2\sin\theta_3 + \sin\theta_1\cos\theta_3 & \cos\theta_1\cos\theta_2 \end{pmatrix}$$

$$\begin{aligned} q_1 &= -\sin(1/2)\theta_1\sin(1/2)\theta_2\sin(1/2)\theta_3 + \cos(1/2)\theta_1\cos(1/2)\theta_2\cos(1/2)\theta_3 \\ q_2 &= +\sin(1/2)\theta_1\cos(1/2)\theta_2\cos(1/2)\theta_3 + \sin(1/2)\theta_2\sin(1/2)\theta_3\cos(1/2)\theta_1 \\ q_3 &= -\sin(1/2)\theta_1\sin(1/2)\theta_3\cos(1/2)\theta_2 + \sin(1/2)\theta_2\cos(1/2)\theta_1\cos(1/2)\theta_3 \\ q_4 &= +\sin(1/2)\theta_1\sin(1/2)\theta_2\cos(1/2)\theta_3 + \sin(1/2)\theta_3\cos(1/2)\theta_1\cos(1/2)\theta_2 \end{aligned}$$

$$\theta_1 = \tan^{-1}\left(\frac{M_{23}}{M_{33}}\right)$$

$$\theta_2 = \tan^{-1}\left(\frac{M_{13}}{\sqrt{1 - M_{13}^2}}\right)$$

$$\theta_3 = \tan^{-1}\left(\frac{M_{12}}{M_{11}}\right)$$

NOTE: This Euler sequence of rotations is used in GNC software for transformations and representations for the attitude-direction indicators. It is also used in the analysis of aeronautical problems and is referred to as the standard *roll-pitch-yaw* sequence.

A.2 Axis Rotation Sequence: 1, 3, 2

$M = M(X(\theta_1), Z(\theta_2), Y(\theta_3)) = XZY$

$$M = \begin{pmatrix} \cos\theta_2\cos\theta_3 & -\sin\theta_2 & \cos\theta_2\sin\theta_3 \\ \cos\theta_1\sin\theta_2\cos\theta_3 + \sin\theta_1\sin\theta_3 & \cos\theta_1\cos\theta_2 & \cos\theta_1\sin\theta_2\sin\theta_3 - \sin\theta_1\cos\theta_3 \\ \sin\theta_1\sin\theta_2\cos\theta_3 - \cos\theta_1\sin\theta_3 & \sin\theta_1\cos\theta_2 & \sin\theta_1\sin\theta_2\sin\theta_3 + \cos\theta_1\cos\theta_3 \end{pmatrix}$$

$$q_1 = +\sin(1/2)\theta_1\sin(1/2)\theta_2\sin(1/2)\theta_3 + \cos(1/2)\theta_1\cos(1/2)\theta_2\cos(1/2)\theta_3$$
$$q_2 = +\sin(1/2)\theta_1\cos(1/2)\theta_2\cos(1/2)\theta_3 - \sin(1/2)\theta_2\sin(1/2)\theta_3\cos(1/2)\theta_1$$
$$q_3 = -\sin(1/2)\theta_1\sin(1/2)\theta_2\cos(1/2)\theta_3 + \sin(1/2)\theta_3\cos(1/2)\theta_1\cos(1/2)\theta_2$$
$$q_4 = +\sin(1/2)\theta_1\sin(1/2)\theta_3\cos(1/2)\theta_2 + \sin(1/2)\theta_2\cos(1/2)\theta_1\cos(1/2)\theta_3$$

$$\theta_1 = \tan^{-1}\left(\frac{M_{32}}{M_{22}}\right)$$

$$\theta_2 = \tan^{1}\left(\frac{-M_{12}}{\sqrt{1 - M_{12}^2}}\right)$$

$$\theta_3 = \tan^{-1}\left(\frac{M_{13}}{M_{11}}\right)$$

A.3 Axis Rotation Sequence: 1, 2, 1

$M = M(X(\theta_1), Y(\theta_2), X(\theta_3)) = XYX$

$$M = \begin{pmatrix} \cos\theta_2 & \sin\theta_2\sin\theta_3 & \sin\theta_2\cos\theta_3 \\ \sin\theta_1\sin\theta_2 & \cos\theta_1\cos\theta_3 - \sin\theta_1\cos\theta_2\sin\theta_3 & -\cos\theta_1\sin\theta_3 - \sin\theta_1\cos\theta_2\cos\theta_3 \\ -\cos\theta_1\sin\theta_2 & \sin\theta_1\cos\theta_3 + \cos\theta_1\cos\theta_2\sin\theta_3 & -\sin\theta_1\sin\theta_3 + \cos\theta_1\cos\theta_2\cos\theta_3 \end{pmatrix}$$

$$
\begin{aligned}
q_1 &= +\cos(1/2)\theta_2 \cos((1/2)(\theta_1 + \theta_3)) \\
q_2 &= +\cos(1/2)\theta_2 \sin((1/2)(\theta_1 + \theta_3)) \\
q_3 &= +\sin(1/2)\theta_2 \cos((1/2)(\theta_1 - \theta_3)) \\
q_4 &= +\sin(1/2)\theta_2 \sin((1/2)(\theta_1 - \theta_3))
\end{aligned}
$$

$$
\begin{aligned}
\theta_1 &= \tan^{-1}\left(\frac{-M_{21}}{M_{31}}\right) \\
\theta_2 &= \tan^{-1}\left(\frac{\sqrt{1 - M_{11}^2}}{M_{11}}\right) \\
\theta_3 &= \tan^{-1}\left(\frac{M_{12}}{M_{13}}\right)
\end{aligned}
$$

A.4 Axis Rotation Sequence: 1, 3, 1

$M = M(X(\theta_1), Z(\theta_2), X(\theta_3)) = XZX$

$$
M = \begin{pmatrix}
\cos\theta_2 & -\sin\theta_2\cos\theta_3 & \sin\theta_2\sin\theta_3 \\
\cos\theta_1\sin\theta_2 & \cos\theta_1\cos\theta_2\cos\theta_3 - \sin\theta_1\sin\theta_3 & -\cos\theta_1\cos\theta_2\sin\theta_3 - \sin\theta_1\cos\theta_3 \\
\sin\theta_1\sin\theta_2 & \sin\theta_1\cos\theta_2\cos\theta_3 + \cos\theta_1\sin\theta_3 & -\sin\theta_1\cos\theta_2\sin\theta_3 + \cos\theta_1\cos\theta_3
\end{pmatrix}
$$

$$
\begin{aligned}
q_1 &= +\cos(1/2)\theta_2 \cos((1/2)(\theta_1 + \theta_3)) \\
q_2 &= +\cos(1/2)\theta_2 \sin((1/2)(\theta_1 + \theta_3)) \\
q_3 &= -\sin(1/2)\theta_2 \sin((1/2)(\theta_1 - \theta_3)) \\
q_4 &= +\sin(1/2)\theta_2 \cos((1/2)(\theta_1 - \theta_3))
\end{aligned}
$$

$$
\begin{aligned}
\theta_1 &= \tan^{-1}\left(\frac{M_{31}}{M_{21}}\right) \\
\theta_2 &= \tan^{-1}\left(\frac{\sqrt{1 - M_{11}^2}}{M_{11}}\right) \\
\theta_3 &= \tan^{-1}\left(\frac{-M_{13}}{M_{12}}\right)
\end{aligned}
$$

A.5 Axis Rotation Sequence: 2, 1, 3

$M = M(Y(\theta_1), X(\theta_2), Z(\theta_3)) = YXZ$

$$M = \begin{pmatrix} \sin\theta_1\sin\theta_2\sin\theta_3 + \cos\theta_1\cos\theta_3 & \sin\theta_1\sin\theta_2\cos\theta_3 - \cos\theta_1\sin\theta_3 & \sin\theta_1\cos\theta_2 \\ \cos\theta_2\sin\theta_3 & \cos\theta_2\cos\theta_3 & -\sin\theta_2 \\ \cos\theta_1\sin\theta_2\sin\theta_3 - \sin\theta_1\cos\theta_3 & \cos\theta_1\sin\theta_2\cos\theta_3 + \sin\theta_1\sin\theta_3 & \cos\theta_1\cos\theta_2 \end{pmatrix}$$

$$q_1 = +\sin(1/2)\theta_1\sin(1/2)\theta_2\sin(1/2)\theta_3 + \cos(1/2)\theta_1\cos(1/2)\theta_2\cos(1/2)\theta_3$$
$$q_2 = +\sin(1/2)\theta_1\sin(1/2)\theta_3\cos(1/2)\theta_2 + \sin(1/2)\theta_2\cos(1/2)\theta_1\cos(1/2)\theta_3$$
$$q_3 = +\sin(1/2)\theta_1\cos(1/2)\theta_2\cos(1/2)\theta_3 - \sin(1/2)\theta_2\sin(1/2)\theta_3\cos(1/2)\theta_1$$
$$q_4 = -\sin(1/2)\theta_1\sin(1/2)\theta_2\cos(1/2)\theta_3 + \sin(1/2)\theta_3\cos(1/2)\theta_1\cos(1/2)\theta_2$$

$$\theta_1 = \tan^{-1}\left(\frac{M_{31}}{M_{33}}\right)$$
$$\theta_2 = \tan^{-1}\left(\frac{-M_{23}}{\sqrt{1 - M_{23}^2}}\right)$$
$$\theta_3 = \tan^{-1}\left(\frac{M_{21}}{M_{22}}\right)$$

A.6 Axis Rotation Sequence: 2, 3, 1

$M = M(Y(\theta_1), Z(\theta_2), X(\theta_3)) = YZX$

$$M = \begin{pmatrix} \cos\theta_1\cos\theta_2 & -\cos\theta_1\sin\theta_2\cos\theta_3 + \sin\theta_1\sin\theta_3 & \cos\theta_1\sin\theta_2\sin\theta_3 + \sin\theta_1\cos\theta_3 \\ \sin\theta_2 & \cos\theta_2\cos\theta_3 & -\cos\theta_2\sin\theta_3 \\ -\sin\theta_1\cos\theta_2 & \sin\theta_1\sin\theta_2\cos\theta_3 + \cos\theta_1\sin\theta_3 & -\sin\theta_1\sin\theta_2\sin\theta_3 + \cos\theta_1\cos\theta_3 \end{pmatrix}$$

$$q_1 = -\sin(1/2)\theta_1\sin(1/2)\theta_2\sin(1/2)\theta_3 + \cos(1/2)\theta_1\cos(1/2)\theta_2\cos(1/2)\theta_3$$
$$q_2 = +\sin(1/2)\theta_1\sin(1/2)\theta_2\cos(1/2)\theta_3 + \sin(1/2)\theta_3\cos(1/2)\theta_1\cos(1/2)\theta_2$$
$$q_3 = +\sin(1/2)\theta_1\cos(1/2)\theta_2\cos(1/2)\theta_3 + \sin(1/2)\theta_2\sin(1/2)\theta_3\cos(1/2)\theta_1$$
$$q_4 = -\sin(1/2)\theta_1\sin(1/2)\theta_3\cos(1/2)\theta_2 + \sin(1/2)\theta_2\cos(1/2)\theta_1\cos(1/2)\theta_3$$

$$\theta_1 = \tan^{-1}\left(\frac{-M_{31}}{M_{11}}\right)$$

$$\theta_2 = \tan^{-1}\left(\frac{M_{21}}{\sqrt{1 - M_{21}^2}}\right)$$

$$\theta_3 = \tan^{-1}\left(\frac{-M_{23}}{M_{22}}\right)$$

A.7 Axis Rotation Sequence: 2, 1, 2

$M = M(Y(\theta_1), X(\theta_2), Y(\theta_3)) = YXY$

$$M = \begin{pmatrix} -\sin\theta_1\cos\theta_2\sin\theta_3 + \cos\theta_1\cos\theta_3 & \sin\theta_1\sin\theta_2 & \sin\theta_1\cos\theta_2\cos\theta_3 + \cos\theta_1\sin\theta_3 \\ \sin\theta_2\sin\theta_3 & \cos\theta_2 & -\sin\theta_2\cos\theta_3 \\ -\cos\theta_1\cos\theta_2\sin\theta_3 - \sin\theta_1\cos\theta_3 & \cos\theta_1\sin\theta_2 & \cos\theta_1\cos\theta_2\cos\theta_3 - \sin\theta_1\sin\theta_3 \end{pmatrix}$$

$$\begin{aligned} q_1 &= +\cos(1/2)\theta_2\cos((1/2)(\theta_1 + \theta_3)) \\ q_2 &= +\sin(1/2)\theta_2\cos((1/2)(\theta_1 - \theta_3)) \\ q_3 &= +\cos(1/2)\theta_2\sin((1/2)(\theta_1 + \theta_3)) \\ q_4 &= -\sin(1/2)\theta_2\sin((1/2)(\theta_1 - \theta_3)) \end{aligned}$$

$$\theta_1 = \tan^{-1}\left(\frac{M_{12}}{M_{32}}\right)$$

$$\theta_2 = \tan^{-1}\left(\frac{\sqrt{1 - M_{22}^2}}{M_{22}}\right)$$

$$\theta_3 = \tan^{-1}\left(\frac{M_{21}}{-M_{23}}\right)$$

A.8 Axis Rotation Sequence: 2, 3, 2

$M = M(Y(\theta_1), Z(\theta_2), Y(\theta_3)) = YZY$

$$M = \begin{pmatrix} \cos\theta_1\cos\theta_2\cos\theta_3 - \sin\theta_1\sin\theta_3 & -\cos\theta_1\sin\theta_2 & \cos\theta_1\cos\theta_2\sin\theta_3 + \sin\theta_1\cos\theta_3 \\ \sin\theta_2\cos\theta_3 & \cos\theta_2 & \sin\theta_2\sin\theta_3 \\ -\sin\theta_1\cos\theta_2\cos\theta_3 - \cos\theta_1\sin\theta_3 & \sin\theta_1\sin\theta_2 & -\sin\theta_1\cos\theta_2\sin\theta_3 + \cos\theta_1\cos\theta_3 \end{pmatrix}$$

$$\begin{aligned} q_1 &= +\cos(1/2)\theta_2\cos((1/2)(\theta_1 + \theta_3)) \\ q_2 &= +\sin(1/2)\theta_2\sin((1/2)(\theta_1 - \theta_3)) \end{aligned}$$

$$q_3 = +\cos(1/2)\theta_2 \sin((1/2)(\theta_1 + \theta_3))$$
$$q_4 = +\sin(1/2)\theta_2 \cos((1/2)(\theta_1 - \theta_3))$$

$$\theta_1 = \tan^{-1}\left(\frac{M_{32}}{-M_{12}}\right)$$

$$\theta_2 = \tan^{-1}\left(\frac{\sqrt{1 - M_{22}^2}}{M_{22}}\right)$$

$$\theta_3 = \tan^{-1}\left(\frac{M_{23}}{M_{21}}\right)$$

A.9 Axis Rotation Sequence: 3, 1, 2

$M = M(Z(\theta_1), X(\theta_2), Y(\theta_3)) = ZXY$

$$M = \begin{pmatrix} -\sin\theta_1 \sin\theta_2 \sin\theta_3 + \cos\theta_1 \cos\theta_3 & -\sin\theta_1 \cos\theta_2 & \sin\theta_1 \sin\theta_2 \cos\theta_3 + \cos\theta_1 \sin\theta_3 \\ \cos\theta_1 \sin\theta_2 \sin\theta_3 + \sin\theta_1 \cos\theta_3 & \cos\theta_1 \cos\theta_2 & -\cos\theta_1 \sin\theta_2 \cos\theta_3 + \sin\theta_1 \sin\theta_3 \\ -\cos\theta_2 \sin\theta_3 & \sin\theta_2 & \cos\theta_2 \cos\theta_3 \end{pmatrix}$$

$$q_1 = -\sin(1/2)\theta_1 \sin(1/2)\theta_2 \sin(1/2)\theta_3 + \cos(1/2)\theta_1 \cos(1/2)\theta_2 \cos(1/2)\theta_3$$
$$q_2 = -\sin(1/2)\theta_1 \sin(1/2)\theta_3 \cos(1/2)\theta_2 + \sin(1/2)\theta_2 \cos(1/2)\theta_1 \cos(1/2)\theta_3$$
$$q_3 = +\sin(1/2)\theta_1 \sin(1/2)\theta_2 \cos(1/2)\theta_3 + \sin(1/2)\theta_3 \cos(1/2)\theta_1 \cos(1/2)\theta_2$$
$$q_4 = +\sin(1/2)\theta_1 \cos(1/2)\theta_2 \cos(1/2)\theta_3 + \sin(1/2)\theta_2 \sin(1/2)\theta_3 \cos(1/2)\theta_1$$

$$\theta_1 = \tan^{-1}\left(\frac{-M_{12}}{M_{22}}\right)$$

$$\theta_2 = \tan^{-1}\left(\frac{M_{32}}{\sqrt{1 - M_{32}^2}}\right)$$

$$\theta_3 = \tan^{-1}\left(\frac{-M_{31}}{M_{33}}\right)$$

A.10 Axis Rotation Sequence: 3, 2, 1

$M = M(Z(\theta_1), Y(\theta_2), X(\theta_3)) = ZYX$

$$M = \begin{pmatrix} \cos\theta_1 \cos\theta_2 & \cos\theta_1 \sin\theta_2 \sin\theta_3 - \sin\theta_1 \cos\theta_3 & \cos\theta_1 \sin\theta_2 \cos\theta_3 + \sin\theta_1 \sin\theta_3 \\ \sin\theta_1 \cos\theta_2 & \sin\theta_1 \sin\theta_2 \sin\theta_3 + \cos\theta_1 \cos\theta_3 & \sin\theta_1 \sin\theta_2 \cos\theta_3 - \cos\theta_1 \sin\theta_3 \\ -\sin\theta_2 & \cos\theta_2 \sin\theta_3 & \cos\theta_2 \cos\theta_3 \end{pmatrix}$$

$$q_1 = +\sin(1/2)\theta_1 \sin(1/2)\theta_2 \sin(1/2)\theta_3 + \cos(1/2)\theta_1 \cos(1/2)\theta_2 \cos(1/2)\theta_3$$
$$q_2 = -\sin(1/2)\theta_1 \sin(1/2)\theta_2 \cos(1/2)\theta_3 + \sin(1/2)\theta_3 \cos(1/2)\theta_1 \cos(1/2)\theta_2$$
$$q_3 = +\sin(1/2)\theta_1 \sin(1/2)\theta_3 \cos(1/2)\theta_2 + \sin(1/2)\theta_2 \cos(1/2)\theta_1 \cos(1/2)\theta_3$$
$$q_4 = +\sin(1/2)\theta_1 \cos(1/2)\theta_2 \cos(1/2)\theta_3 - \sin(1/2)\theta_2 \sin(1/2)\theta_3 \cos(1/2)\theta_1$$

$$\theta_1 = \tan^{-1}\left(\frac{M_{21}}{M_{11}}\right)$$
$$\theta_2 = \tan^{-1}\left(\frac{-M_{31}}{\sqrt{1 - M_{31}^2}}\right)$$
$$\theta_3 = \tan^{-1}\left(\frac{M_{32}}{M_{33}}\right)$$

NOTE: This is the aeronautical standard *yaw-pitch-roll* Euler sequence. It is used because pilots and engineers can easily visualize the aircraft pitch and roll orientation angles relative to the horizontal plane.

A.11 Axis Rotation Sequence: 3, 1, 3

$M = M(Z(\theta_1), X(\theta_2), Z(\theta_3)) = ZXZ$

$$M = \begin{pmatrix} -\sin\theta_1\cos\theta_2\sin\theta_3 + \cos\theta_1\cos\theta_3 & -\sin\theta_1\cos\theta_2\cos\theta_3 - \cos\theta_1\sin\theta_3 & \sin\theta_1\sin\theta_2 \\ \cos\theta_1\cos\theta_2\sin\theta_3 + \sin\theta_1\cos\theta_3 & \cos\theta_1\cos\theta_2\cos\theta_3 - \sin\theta_1\sin\theta_3 & -\cos\theta_1\sin\theta_2 \\ \sin\theta_2\sin\theta_3 & \sin\theta_2\cos\theta_3 & \cos\theta_2 \end{pmatrix}$$

$$q_1 = +\cos(1/2)\theta_2 \cos((1/2)(\theta_1 + \theta_3))$$
$$q_2 = +\sin(1/2)\theta_2 \cos((1/2)(\theta_1 - \theta_3))$$
$$q_3 = +\sin(1/2)\theta_2 \sin((1/2)(\theta_1 - \theta_3))$$
$$q_4 = +\cos(1/2)\theta_2 \sin((1/2)(\theta_1 + \theta_3))$$

$$\theta_1 = \tan^{-1}\left(\frac{M_{13}}{-M_{23}}\right)$$
$$\theta_2 = \tan^{-1}\left(\frac{\sqrt{1 - M_{33}^2}}{M_{33}}\right)$$
$$\theta_3 = \tan^{-1}\left(\frac{M_{31}}{M_{32}}\right)$$

NOTE: This Euler sequence is referred to by the author as the classical Euler angle set and is used extensively in problems of astronomy and astrodynamics. The geometric orientation of orbits in space is defined using the standard: *right ascension, orbital inclination, and argument of perigee* angles in this 3-1-3 sequence.

A.12 Axis Rotation Sequence: 3, 2, 3

$M = M(Z(\theta_1), Y(\theta_2), Z(\theta_3)) = ZYZ$

$$M = \begin{pmatrix} \cos\theta_1\cos\theta_2\cos\theta_3 - \sin\theta_1\sin\theta_3 & -\cos\theta_1\cos\theta_2\sin\theta_3 - \sin\theta_1\cos\theta_3 & \cos\theta_1\sin\theta_2 \\ \sin\theta_1\cos\theta_2\cos\theta_3 + \cos\theta_1\sin\theta_3 & -\sin\theta_1\cos\theta_2\sin\theta_3 + \cos\theta_1\cos\theta_3 & \sin\theta_1\sin\theta_2 \\ -\sin\theta_2\cos\theta_3 & \sin\theta_2\sin\theta_3 & \cos\theta_2 \end{pmatrix}$$

$$\begin{aligned} q_1 &= +\cos(1/2)\theta_2\cos((1/2)(\theta_1+\theta_3)) \\ q_2 &= -\sin(1/2)\theta_2\sin((1/2)(\theta_1-\theta_3)) \\ q_3 &= +\sin(1/2)\theta_2\cos((1/2)(\theta_1-\theta_3)) \\ q_4 &= +\cos(1/2)\theta_2\sin((1/2)(\theta_1+\theta_3)) \end{aligned}$$

$$\theta_1 = \tan^{-1}\left(\frac{M_{23}}{M_{13}}\right)$$

$$\theta_2 = \tan^{-1}\left(\frac{\sqrt{1-M_{33}^2}}{M_{33}}\right)$$

$$\theta_3 = \tan^{-1}\left(\frac{M_{32}}{-M_{31}}\right)$$

Appendix B: C-W State Transition Matrix for LVLH Relative Motion

The following is a presentation of the C-W solution for the propagation of the LVLH relative motion of a space vehicle as seen from a space vehicle in circular orbit. This solution was derived by William Jackson at NASA Johnson Space Center[39] and includes the constant accelerations that may be acting on the rendezvous space vehicle. The propagation equation is described in Eq. (2.143) of Sec. 2.4.2. The elements of the 6×9, T matrix are written here with six matrix partitions that are used in rendezvous and guidance applications as presented in Sec. 2.4.4. This notation also simplifies the understanding of details of the rendezvous GNC applications. The T matrix is formed by partitioning into 3×3 matrices, so that we can write Eq. (2.143) as

$$\bar{\boldsymbol{x}}(t) = \begin{pmatrix} T_{11} & T_{12} & T_{13} \\ T_{21} & T_{22} & T_{23} \end{pmatrix} \bar{\boldsymbol{x}}(t_o) \tag{B.1}$$

The propagation time is $\Delta t = (t - t_o)$ and ω_o is the orbital rate as given in Eq. (2.151). Also let $\theta_t = \omega_o \Delta t$, and based on the partitions in Eq. (B.1), we have for position transition from initial position,

$$T_{11} = \begin{pmatrix} 1 & 0 & 6(\theta_t - \sin\theta_t) \\ 0 & \cos\theta_t & 0 \\ 0 & 0 & (4 - 3\cos\theta_t) \end{pmatrix}$$

for position transition from initial velocity,

$$T_{12} = \begin{pmatrix} (4\sin\theta_t/\omega_o) - 3\Delta t & 0 & 2(1 - \cos\theta_t)/\omega_o \\ 0 & \sin\theta_t/\omega_o & 0 \\ -2(1 - \cos\theta_t)/\omega_o & 0 & \sin\theta_t/\omega_o \end{pmatrix}$$

for position transition from the constant acceleration,

$$T_{13} = \begin{pmatrix} (4/\omega_o^2)(1 - \cos\theta_t) - 1.5(\Delta t)^2 & 0 & 2(\theta_t - \sin\theta_t)/{\omega_o}^2 \\ 0 & (1 - \cos\theta_t)/{\omega_o}^2 & 0 \\ -2(\theta_t - \sin\theta_t)/{\omega_o}^2 & 0 & (1 - \cos\theta_t)/{\omega_o}^2 \end{pmatrix}$$

for velocity transition from initial position,

$$T_{21} = \begin{pmatrix} 0 & 0 & 6\omega_o(1 - \cos\theta_t) \\ 0 & -\omega_o \sin\theta_t & 0 \\ 0 & 0 & 3\omega_o \sin\theta_t \end{pmatrix}$$

for velocity transition from initial velocity,

$$T_{22} = \begin{pmatrix} 4\cos\theta_t - 3 & 0 & 2\sin\theta_t \\ 0 & \cos\theta_t & 0 \\ -2\sin\theta_t & 0 & \cos\theta_t \end{pmatrix}$$

and finally for velocity transition from the constant acceleration,

$$T_{23} = \begin{pmatrix} (4/\omega_o)\sin\theta_t - 3\Delta t & 0 & 2(1 - \cos\theta_t)/\omega_o \\ 0 & (\sin\theta_t)/\omega_o & 0 \\ -2(1 - \cos\theta_t)/\omega_o & 0 & (\sin\theta_t)/\omega_o \end{pmatrix} \tag{B.2}$$

Appendix C: Integral Lists for Computer Simulation Algorithms

In this appendix integral lists are presented that are used for the numerical solution of many aerospace problems. Computer algorithms have been developed, as those described in Appendix D, using first-order solutions for the differential equations of motion. These solutions can be generalized using an array of state parameters, $y_1, y_2, y_3, \ldots, y_n$, and an array of their first derivatives, $\dot{y}_1, \dot{y}_2, \dot{y}_3, \ldots, \dot{y}_n$. The numerical solution is characterized by the stepwise integral:

$$y_i(t + \Delta t) = y_i(t) + \int_t^{t+\Delta t} \dot{y}_i(t)\,\mathrm{d}t \qquad \text{for } i = 1, n \tag{C.1}$$

This integral is performed in Δt steps until the end boundary condition is reached, for example, until the state parameter achieves a desired value or until the stop time is reached. The following integral lists refer to the indicated solutions of Eq. (C.1).

C.1 Three-DOF Motion Simulation

	y_i	$\dot{y}_i$
Position State		
1.	x_1	$\dot{x}_1$
2.	x_2	$\dot{x}_2$
3.	x_3	$\dot{x}_3$
Velocity State		Acceleration from EOM
4.	$\dot{x}_1$	$\ddot{x}_1$
5.	$\dot{x}_2$	$\ddot{x}_2$
6.	$\dot{x}_3$	$\ddot{x}_3$

C.2 Four-DOF Airframe Motion Simulation

	$\underbrace{y_i}$	$\underbrace{\dot{y}_i}$
Position State		
1.	x_1	$\dot{x}_1$
2.	x_2	$\dot{x}_2$
3.	x_3	$\dot{x}_3$
Velocity State		Acceleration from EOM
4.	$\dot{x}_1$	$\ddot{x}_1$
5.	$\dot{x}_2$	$\ddot{x}_2$
6.	$\dot{x}_3$	$\ddot{x}_3$
Pitch Attitude		
7.	θ	$\dot{\theta}$
Pitch Rate		Pitch Acceleration from EOM
8.	$\dot{\theta}$	$\ddot{\theta}$
Vehicle Weight		Weight Flow Rate
9.	W	$\dot{W}$

C.3 Six-DOF Airframe Motion Simulation

	$\underbrace{y_i}$	$\underbrace{\dot{y}_i}$
Position State		
1.	x_1	$\dot{x}_1$
2.	x_2	$\dot{x}_2$
3.	x_3	$\dot{x}_3$
Velocity State		Acceleration from EOM
4.	$\dot{x}_1$	$\ddot{x}_1$
5.	$\dot{x}_2$	$\ddot{x}_2$
6.	$\dot{x}_3$	$\ddot{x}_3$
Attitude Quaternion		Quaternion Rates
7.	q_1	$\dot{q}_1$
8.	q_2	$\dot{q}_2$
9.	q_3	$\dot{q}_3$
10.	q_4	$\dot{q}_4$
Body Axis Rotation Rates		Rotational Accelerations from EOM
11.	ω_1	$\dot{\omega}_1$
12.	ω_2	$\dot{\omega}_2$
13.	ω_3	$\dot{\omega}_3$
Vehicle Weight		Weight Flow Rate
14.	W	$\dot{W}$
Total Delta-V		From Contact Force Accelerations
15.	ΔV	$\dot{V}$

Delta-V_V

16. $\triangle V_V$ $\dot{V}_V$

Delta-$V_{\perp V}$

17. $\triangle V_{\perp V}$ $\dot{V}_{\perp V}$

Delta-V_g

18. $\triangle V_g$ $\dot{V}_g$

Delta-$V_{\perp g}$

19. $\triangle V_{\perp g}$ $\dot{V}_{\perp g}$

Appendix D: Numerical Solution Methods for Differential Equations: Computer Simulation Algorithms

In this appendix suggested logic flow methods are presented for the solutions of differential equations of motion. Each particular aerospace problem, which requires the numeric solutions of the differential equations of motion, calls for the application of the methods best suited for that problem. The methods shown in this appendix find application when simulating sensors, navigation, guidance, and control functions of the aerospace vehicles. In these simulations very small cycle times are required to capture the rapidly changing control functions and the rotational body dynamics of the vehicle. Although the translational motion parameters are changing at much slower rates, the GNC cycle times dictate the solution methods to simulate the motion of the vehicle. Many computer algorithms have been developed from these simple example applications.

D.1 Cauchy-Euler

This is the simplest numerical solution method for n first-order differential equations of motion. This solution method requires one evaluation of the equations of motion per time step.

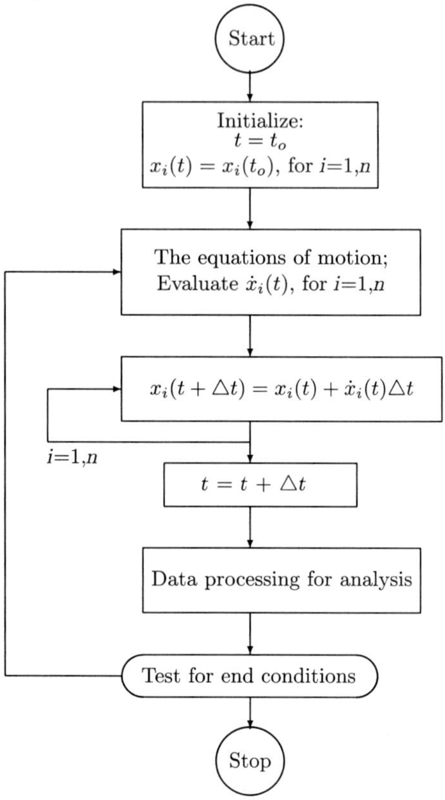

D.2 Midpoint Cauchy-Euler

This solution method requires two evaluations of the equations of motion per time step.

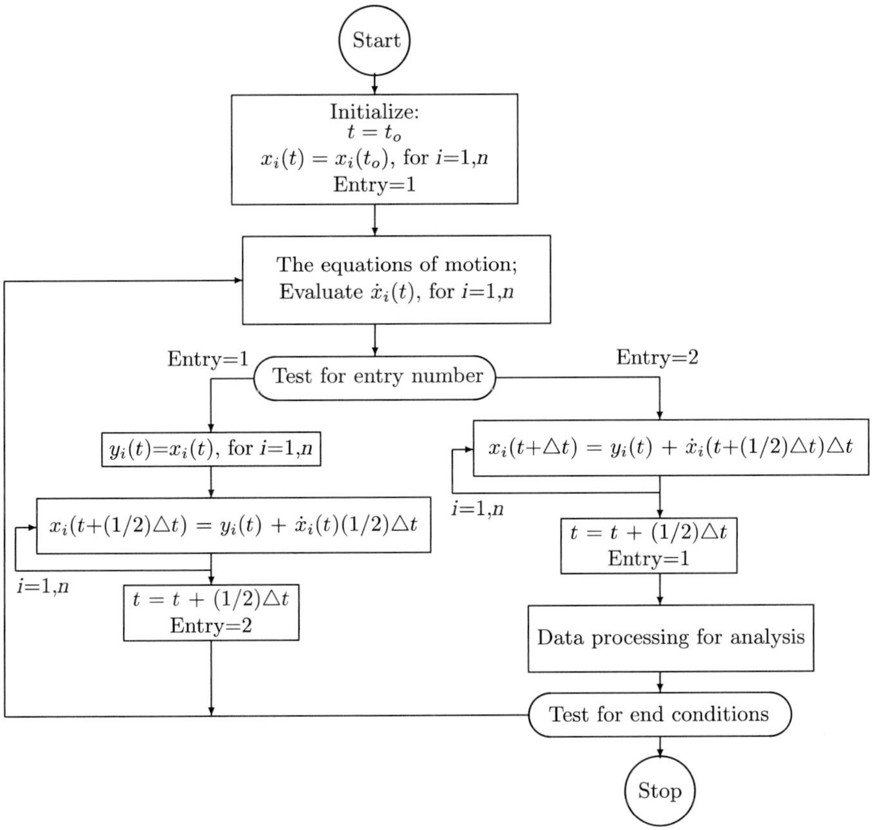

References

[1]Lorentz, H. A., Einstein, A., Minkowski, H., and Weyl, H., *The Principle of Relativity*, Dover, New York.

[2]Adler, R., Bazin, M., and Schiffer, M., *Introduction to General Relativity*, McGraw-Hill, New York, 1965.

[3]Byron, F. W., Jr., and Fuller, R. W., *Mathematics of Classical and Quantum Physics*, Dover, New York, 1992.

[4]Lass, H., *Vector and Tensor Analysis*, McGraw-Hill, New York, 1950.

[5]Data for Astronomy, Space Sciences, Geodesy, Surveying, Navigation and Other Applications, *The Astronomical Almanac of the Year 2005*, U.S. Governmemt Printing Office, Washington, DC, and The Stationary Office, London, Sept. 2003.

[6]Perlis, S., *Theory of Matrices*, Addison-Wesley, Reading, MA, 1952.

[7]Bond, V. R., "Precessional, Nutational, and Earth Rotational Transformations Based upon the Standard Epoch J2000," NASA JSC-22535, June 1987.

[8]Vallado, D. A., *Fundamentals of Astrodynamics and Applications*, 2nd ed., Microcosm Press, El Segundo, CA, and Kluwer Academic Publishers, Dordrecht, The Netherlands, 2001.

[9]Corben, H. C., and Stehle, P., *Classical Mechanics*, 2nd ed., Wiley New York, 1960.

[10]VanderVoort, R., "Notes on the Carousel IV Math Model," Apollo Engineering, March 1971.

[11]Whittaker, E. T., *A Treatise on the Analytical Dynamics of Particles and Rigid Bodies*, 4th ed., Cambridge Univ. Press, 1959.

[12]Hamilton, William Rowan, *Elements of Quaternions*, 3rd ed., Chelsea, New York, 1969.

[13]Goldstein, H., *Classical Mechanics*, Addison-Wesley, Cambridge, MA, 1956.

[14]Bean, W. C., "Operational Use of Quaternions in the Description of Rigid Body Motion," NASA MSC Internal Note 69-FM-237, Oct. 1969.

[15]Ickes, B. P., "A New Method for Performing Digital Control Attitude Computations Using Quaternions," *AIAA Journal*, Jan. 1, 1970, pp. 13–17.

[16]Carroll, J. V., "The Notation and Use of Quaternions for Shuttle Ascent Steering," Charles Stark Draper Lab., SSV Memo 10C-75-47, Cambridge, MA, Nov. 1975.

[17]Klumpp, A. R., "Singularity-Free Extraction of a Quaternion from a Direction-Cosine Matrix," Charles Stark Draper Lab., Shuttle Memo 10E-76-4, TD 3.3, Cambridge, MA, Jan. 1976.

[18]Henderson, D. M., "Recursive Relationships for Body Axis Rotation Rates," *Journal of Guidance, Control, and Dynamics*, Vol. 6, No. 4, 1983, pp. 318–320.

[19]Etkin, B., *Dynamics of Atmospheric Flight*, Wiley, New York, 1972.

[20]Newton, Isaac, *Mathematical Principles of Natural Philosophy and the System of the World*, translated by F. Cajori, Univ. of California Press, Berkeley, CA, 1934.

[21]Symon, K. R., *Mechanics*, 3rd ed., Addison-Wesley, Reading, MA, 1971.

[22]Ehricke, K. A., *Space Flight, II. Dynamics*, Van Nostrand, Princeton, NJ, 1962.

[23]Cockrell, B. F., "Space Shuttle Astrodynamical Constants," NASA JCS-14262, Nov. 1983.

[24]Escobal, P. R., *Methods of Orbit Determination*, Wiley, New York, 1965.

[25]Standish, E. M., Jr., Keesey, M. S. W., and Newhall, X. X., "JPL Development Ephemeris Number 96," Jet Propulsion Lab., California Inst. of Technology, TR 32-1603, Pasadena, CA, Feb. 1976.

[26]Meirovitch, L., *Methods of Analytical Dynamics*, McGraw-Hill, New York, 1970.

[27]Bond, V. R., and Allman, M. C., *Modern Astrodynamics*, Princeton Univ. Press, Princeton, NJ, 1996.

[28]Taff, L. G., *Celestial Mechanics, a Computational Guide for the Practitioner*, Wiley, New York, 1985.

[29]Hildebrand, F. B., *Introduction to Numerical Analysis*, McGraw-Hill, New York, 1956.

[30]Margenau, H., and Murphy, G., *The Mathematics of Physics and Chemistry*, 2nd ed., Van Nostrand, 1956.

[31]Baker, R. M. L., Jr., and Makemson, M. W., *An Introduction to Astrodynamics*, 2nd ed., Academic Press, New York, 1967.

[32]"Natural Environment and Physical Standards for the Apollo Program and Apollo Applications Program," NASA M-DE-8020.00C, July 1969.

[33]Kuethe, A. M., and Chow, C.-Y., *Foundations of Aerodynamics, Bases of Aerodynamic Design*, 4th ed., Wiley, New York, 1986.

[34]Anderson, J. D., Jr., *Introduction to Flight*, 2nd ed., McGraw-Hill, New York, 1985.

[35]Perkins, C. D., and Hage, R. E., *Airplane Performance Stability and Control*, Wiley, New York, 1957.

[36]Hill, G. W., "Researchers in Lunar Theory," *American Journal of Mathematics*, Vol. 1:5–26, 1879.

[37]Clohessy, W. H., and Wiltshire, R. S., "A Terminal Guidance System for Satellite Rendezvous," *Aero/Space Science*, Vol. 27, Sept. 1960, p. 653.

[38]Jackson, W. L., "Relative Motion EOM for Constant Accelerations," NASA/JSC EH2-86M-104, Feb. 1986.

[39]Neily, C. M., and Jackson, W. L., "An Efficient Computational Algorithm for Multi-Burn Orbital Rendezvous Maneuvers," AAS Paper 85-441, Aug. 1985.

[40]Abdi-Moradi, M., and Henderson, D. M., "GENBOD User's Manual; The TRW Mass Properties Program," TRW, Rept. 87:W480.1-6, Houston, TX, 1987.

[41]Faddeeva, V. N., *Computational Methods of Linear Algebra*, Dover, New York, 1959.

[42]Baker, R. M. L., Jr., *Astrodynamics, Applications and Advanced Topics*, Academic Press, New York, 1967.

[43]Dashiell, J. N., "Gravity Gradient Torques on a Rigid Body," TRW, Rept. 67:7252.6-124, Oct. 1967.

[44]Henderson, D. M., "Solution of Euler's Equation of Motion for a Rigid Body Using the Quaternion," AIAA Technical Paper, NASA JSC, March 21, 1978.

[45]MacCullagh, J., *Transactions of the Royal Irish Academy*, Vol. XXII, Parts I and II, 1855.

[46]Lear, W. M., "The Gravitational Acceleration Equations," NASA JSC-22080, April 1986.

[47]Henderson, D. M., "Long-Term Space Station Orbit Propagation Using Perturbation Methods," NASA JSC-32081, May 1988.

[48]Bond, V. R., and Gottlieb, R. G., "A Perturbed Two-Body Element Method Utilizing the Jacobian Integral," NASA JSC-23495, May 1989.

[49]Lear, W. M., "Subroutines to Compute the Gravitational Acceleration," NASA JSC-25469, Feb. 1992.

[50]Spencer, J. L., "Use of a Non-singular Potential," NASA JSC-09661, May 1975.

[51]Gottlieb, R. G., "A Fast Recursive Singularity Free Algorithm for Calculating the First and Second Derivatives of the Geopotential," NASA JSC-23762, July 1990.

[52]Cockrell, B. F., "Space Shuttle Astrodynamical Constants, Revision 1," NASA JSC-14262, Nov. 1983.

[53]*U.S. Standard Atmosphere, 1976*, NOAA, NASA, VSAF, NOAA-S/T 76-1562, Oct. 1976.

[54]Jacchia, L. G., "Thermospheric Temperature, Density, and Composition: New Models," Smithsonian Institution, Astrophysical Observatory, SAO Special Rept. 375, Cambridge, MA, March 1977.

[55]Rubbert, P. E., and Saaris, G. R., "Review and Evaluation of a Three-Dimensional Lifting Potential Flow Analysis for Arbitrary Configurations," AIAA Paper 72–188, 1972.

[56]Anderson, J. D., Jr., *Fundamentals of Aerodynamics*, McGraw-Hill, New York, 1984.

[57]Rudy, D. H., Morris, D. J., Blanchard, D. K., Cook, C. H., and Rubin, S. G., "An Investigation of Several Numerical Procedures for Time-Asymptotic Compressible Navier-Stokes Solutions," NASA SP-347, 1975.

[58]Ley, W., *Rockets, Missiles, and Space Travel*, Viking Press, New York, 1951.

[59]Cowell, P. H., and Crommelin, *Investigation of the Motion of Halley's Comet*, Appendix, Greenwich Observations, 1909.

[60]Szebehely, V., *Theory of Orbits: The Restricted Problem of Three Bodies*, Academic Press, New York, 1967.

[61]Sundmann, K., "Memoire sur le probleme des trois corps," *Acta Mathematika*, 36, 1912.

[62]Brouwer, D., and Clemence, G., *Methods of Celestial Mechanics*, Academic Press, 1961.

[63]McCuskey, S. W., *Introduction to Celestial Mechanics*, Addison-Wesley, 1963.

[64]Kozai, Y., "The Motion of a Close Earth Satellite," *Astronomical Journal*, Vol. 64, No. 1274, 1959.

[65]Ward, J., "Variations in Orbital Elements," Air Force Eastern Test Range (AFSC), Rept. ETR-TR-71-4, Patrick AFB, FL, Aug. 1971.

[66]Hoots, F. R., and Roehrich, R. L., "Models for Propagation of NORAD Element Sets," Aerospace Defense Command, Space Track Rept. No. 3, U.S. Air Force, Dec. 1980.

[67]Brouwer, D., "Solution of the Problem of Artificial Satellite Theory Without Drag," *Astronomical Journal*, No. 64, 1959, pp. 378–397.

[68]Giacaglia, G. "The Influence of High-Order Zonal Harmonics on the Motion of an Artificial Satellite Without Drag," *Astronomical Journal*, No. 69, 1946, pp. 303–308 (see Erratum: *Astronomical Journal* No. 71, 1966, p. 228).

[69]Lyddane, R. H., "Small Eccentricities or Inclinations in the Brouwer Theory of the Artificial Satellite," *Astronomical Journal*, No. 68, 1963, pp. 555–561.

[70]Izak, I., "A Note on Perturbation Theory," *Astronomical Journal*, No. 68, 1963, pp. 559–561.

[71]Frazer, J. B., "On the Motion of an Artificial Earth Satellite," Mitre Corp., Rept. ESD-TR-66-293, Sep. 1966.

[72]Henderson, D. M., "Computation of Orbital Conjunctions and Collision Prediction," AAS Paper AAS99-174, Feb. 1999.

[73]Kimball, G. R., "Euler Matrices," Apollo Guidance Program Section, U.S. Government Memorandum, March 20, 1966.

[74]Henderson, D. M., "Euler Angles, Quaternions and Transformation Matrices—Working Relationships," NASA JSC-12960, July 1977.

[75]Bond, V. R., *The Elimination of Secular Terms from the Differential Equations of the BG14 Element Method*, NASA JSC-24351, May 1990.

[76]Jacchia, L. G., "New Static Models of the Thermosphere and Exosphere with Empirical Temperature Profiles," Smithsonian Institution, Astrophysical Observatory, SAO Special Rept. 313, Cambridge, MA, May 1970.

Index

Supporting Materials

A complete listing of titles in the AIAA Education Series and other AIAA publications is available at http://www.aiaa.org.